sleeping naked is green

sleeping naked is green

how an eco-cynic unplugged her fridge, sold her car, and found love in 366 days

VANESSA FARQUHARSON

Houghton Mifflin Harcourt Publishing Company
Boston New York 2009

www.hmhbooks.com

Library of Congress Cataloging-in-Publication Data

Farquharson, Vanessa.
Sleeping naked is green : how an eco-cynic unplugged her fridge, sold her car, and found love in 366 days / Vanessa Farquharson.
p. cm.
ISBN-13: 978-0-547-07328-6
ISBN-10: 0-547-07328-3
1. Sustainable living. 2. Green movement. 3. Environmental protection—Citizen protection. 4. Farquharson, Vanessa—Homes and haunts. 5. Farquharson, Vanessa—Friends and associates. 6. Journalists—Canada. I. Title.
GE195.7.F37 2009
333.72092—DC22 200853141

Book design by Anne Chalmers
Typefaces: Minion, Meta, Oxtail

Printed in the United States of America
DOC 10 9 8 7 6 5 4 3 2 1

The text stock is 100% postconsumer waste recycled and FSC certified. (Forest Stewardship Council certification ensures the environmentally responsible, socially acceptable, and economically viable use of well-managed forests.)

BIO GAS

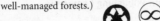

For Grandma Red and G-Dad

contents

introduction

We should respect the earth. We should live responsibly. We should change our light bulbs, try composting, eat local organic food, and build more bike lanes.

When it comes to the green movement, everyone from politicians to musicians talks about what *should* be done. But no one seems to be talking about what all this doing actually entails.

Yes, of course, let's all start composting—but can anybody explain how, exactly, one stores a tub of rotting Chinese take-out, shredded newspaper, and worms in an open-concept kitchen in a seven-hundred-square-foot apartment? Yes, it's important to buy local and organic, but what do you do when your choice at the grocery store is between a pesticide-coated Royal Gala from just a few miles away and an organic Granny Smith flown in from New Zealand? There are so many "Top 10 Ways to Go Green" lists floating around in magazines and on websites and talk shows these days that most people could recite them by heart, starting with the obvious "Switch to Compact Fluorescent Light Bulbs!"

But what of the sickly glow they give off? And what if half the light fixtures in your house are wired for halogens? And what about all that plastic packaging they come wrapped in? So much of the green dialogue comes to a grinding halt with a list of redundant tips; empty reassurances about how easy, fun, and chic it is to be green; or finger-pointing at corporations, governments, and the general public about whose responsibility it is to save the planet and whether we're all doomed.

Enough already. It's time people stopped talking so much and started doing something—anything.

I realized this in February 2007 and decided that doing something, in my case, would begin with a resignation letter. No more sitting in a cubicle, eating cafeteria food, and staring at a computer screen. I was going to quit my job as an arts reporter at the *National Post* and move to an exotic destination like Cambodia or Sri Lanka, where I could work for a charity, preferably an elephant orphanage or some cushier branch of the World Wildlife Fund.

Unfortunately, it didn't take long before I realized that such organizations were more in need of a "Project Manager, experience required," than a "Bitter Ex-Journalist, experience limited to CuteOverload.com." And although I was sure that, with some time, I could manage a project—I had, after all, paid close attention to the first two seasons of *The Apprentice*—it might not have been so easy to convince others of this. So I forgot about animals.

Instead, I turned to food. Maybe I could go work on an organic vegetable farm in northern Ontario, planting, tilling, and harvesting in exchange for room and board. I might not effect massive change, but at least I'd be living as simply as possible, learning about sustainable agri-business and ethical eating.

But even then, to be a farmhand requires knowing the difference between kale and Swiss chard, being willing to trade a morning latte for a shovel, and knowing some basics of soil maintenance—and, well, considering I can barely maintain a rosemary bush on my balcony, I figured I'd only be a burden.

Eventually, I came up with a more practical idea:

What if I didn't quit my job? I was already a reporter at a national newspaper and what more powerful tool for communicating the importance of environmental issues than the media? Even better, the *National Post* is renowned for its decidedly conservative, antigreen bent—we routinely run editorials calling environmentalists "eco-fascists" and the "Green Gestapo" and have one columnist who, despite riding a bike on weekends and avoiding Styrofoam containers at lunch, seems to have made it his life mission to prove Al Gore wrong. If, then, I could somehow convince my editors to give me a column, perhaps just some space on the *Post*'s website in which I could write about this topic, I might be able to make a difference. In fact, providing all went according to plan, maybe I could lay claim to single-handedly greening the most un-green newspaper in the country. I knew nothing about the science of climate change, the technology behind solar panels, or why #2 plastics were superior to #5 plastics in the waste hierarchy, but I did know that I planned on investing in a few tote bags and using public transit more often. And yet, I remained uncertain as to whether I could actually turn these armchair-environmentalist ambitions into meaningful journalism, or at least into something people would actually read.

This is when the proverbial compact fluorescent light bulb above my head lit up. As I lay in bed one night, tossing and turning with the carbon guilt of driving to work and back that day without carpooling, I started to think about the cycle of cynicism and the cycle of hope, both of which I'd just read about in *The Better World Handbook: Small Changes That Make a Big Difference.* The cycle of cynicism goes something like this:

1. Finding out about a problem
2. Wanting to do something to help
3. Not seeing how you can help
4. Not doing anything about it
5. Feeling sad, powerless, angry
6. Deciding that nothing can be done
7. Beginning to shut down
8. Wanting to know less about problems

   (*Repeat until apathy results.*)

(Aside: This is pretty much me, in an eight-step nutshell.)

Then there's the cycle of hope, which, for whatever reason, has two fewer steps (those optimists are always looking for shortcuts), and goes like this:

1. Taking personal responsibility for being a good person
2. Creating a vision of a better world based on your values
3. Seeking out quality information about the world's problems
4. Discovering practical options for actions
5. Acting in line with your values
6. Recognizing you can't do everything
(Repeat until a better world results.)

Then it came to me: if I could do one small thing every day for a year—small changes, like the tote bags, but maybe some big changes too, like overhauling my diet and restricting my consumerism—I'd see which things were easy, which were hard; which ideas would apply only to a single woman like me, living with her cat in the city, and which ones might also apply to a family of four living in a suburb; which changes we should all be doing and which are best left to the hardcore hippies. It's like that Chinese proverb: The journey of a thousand miles begins with a single step. I would take 365 of them.

But I'd need some company. If the *Post* didn't agree to a column, I'd need another outlet. My former colleague Kelly used to keep up a theater blog, and my friend Meghan had just started a nutrition blog—I couldn't even say the word *blog* without feeling as though I'd suddenly joined a *Star Trek* fan club and bought high-waisted Dockers, but it really would be the perfect medium. This meant I could do something every day and then write about it immediately. And, if I kept my skepticism in high gear—making every effort not to be preachy, schmaltzy, or self-righteous—I might attract more readers. The greater the following the blog had, the more committed I'd have to be, both to the supporters and the dissenters. After all, I couldn't quit with everyone watching.

This was my sincere, ambitious, and in hindsight ridiculously naïve line of thinking as I went about setting up the blog, soliciting input, thinking up my first few green changes, and e-mailing the *Post*'s editor in chief about this Pulitzer Prize–winning idea.

A week went by.

He didn't write back.

No problem, I thought, he's probably busy doing important editor business. So, a week later when everything was more or less ready to go, I wrote to our managing editor, who was big into all things Internet-related but still learning that it wasn't cool to use the term "Information Superhighway," and I forwarded along my URL, asking if he could maybe link to it on the *Post*'s homepage or give me some feedback.

Two weeks went by this time, but then he wrote back.

"Why don't you write a film blog?" he said.

Okay, so it was going to be harder than I thought. But still, I rationalized, if I could cultivate a steady readership and build up enough daily hits, my superiors—most of whom had only just figured out what RSS feeds and tag clouds actually were—would surely come knocking at my cubicle at some point down the road, right? And even if they didn't, was that any reason not to go about my green year, showing at the very least my family, friends, and some fellow blogging nerds that being environmentally conscious isn't such a daunting prospect after all? Furthermore, it took me an entire weekend to set up that damn website, so frankly, whether or not anyone was listening, I was going ahead with it for the sake of Mother Earth, and the compost was hitting the wind turbine on March 1, 2007.

Because it felt somewhat like coming out of the environmental closet, the first people to hear about this venture were those who'd have no choice but to support me, people who'd have to love me even if I reeked of hemp oil and started growing dreadlocks from my armpits: my parents.

"It's not like I'll be doing anything crazy," I explained to the two furrowed brows, one tilted head, and a half-opened mouth that formed their combined expression of concern. We were out for dinner, the day before my green challenge began, which as I explained to them would involve saving the planet in 365 simple steps and tracking my progress online.

"Day one will be switching to recycled paper towels," I said. "Day two could be something like not using my electric heating pad anymore. Easy stuff."

Yes, I admitted, there's a chance I may have seen *An Inconvenient Truth* at some point in the past few weeks, and yes, I may have been influenced by that computer-generated drowning polar bear. But the point of this is to stay far away from politics, to prove that being environmentally friendly doesn't require protest demonstrations or wearing Guatemalan pants. After all, I pointed out, they knew more than anyone that their firstborn daughter was the kind of girl who likes her heat turned up high and her incandescent lights dimmed low, who enjoys blowing her month's savings on a bottle of pink Veuve Clicquot and pairing it with back-to-back reruns of *America's Next Top Model.* They knew full well that I held a British passport and thus kept my sense of humor as dry as week-old lint—I wasn't a serious eco-warrior who threw around words like *permaculture* and refused to shut up about the health benefits of wheatgrass. Please.

"So you see," I said, "if I can be good to the environment without compromising my need for flattering lighting, overpriced champagne, and reality TV, then anyone can do it. Right? Tell me I'm right."

"My little activist daughter," cooed my mother eventually, with 90 percent encouragement and 10 percent sarcasm, over the remains of her filet mignon. I wondered if it was factory-farmed meat, then wondered if I really cared.

"And what's day three?" my father asked.

"Day three?" I said. "Well, I haven't planned that far off . . ."

sleeping naked is green

march

1	Switch to recycled paper towels
2	No more electric heating pad
3	Ban all Styrofoam
4	Switch to an eco-friendly toothbrush
5	Turn down my thermostat
6	No more bottled water
7	Switch to an organic conditioner
8	Use natural (pumice or felt) lint removers rather than tape
9	Eat locally (within the United States and Canada)
10	Switch to electronic billing
11	Check the tire pressure
12	Don't buy a microwave
13	Switch to a natural, biodegradable hand wash with a recyclable container
14	Switch to phosphate-free dish detergent
15	Forgo electronic gym equipment
16	Use tote bags
17	Switch to recycled toilet paper
18	Cancel cable
19	Switch to corn-based, biodegradable cat litter
20	Use handkerchiefs instead of tissues
21	Use natural, organic body lotion
22	No more aerosol cans
23	Use natural glass cleaners
24	Sign up with local Freecycle network
25	Turn off all lights before leaving home
26	Switch to natural body wash
27	Use chemical-free, reusable static-cling sheets in the dryer
28	Unplug electronics and appliances when not in use
29	No more disposable cutlery and plates
30	No car on the weekends
31	Put away humidifier

MARCH 1, DAY 1

Switch to recycled paper towels

Asinine. A word I learned in Ms. Carrier's eleventh-grade English class and one that I think works perfectly in the following sentence:

Thinking up 365 ways to green my life and writing about it every single day for an entire year probably qualifies as the most asinine idea I've come up with in twenty-eight years of otherwise uneventful neurological activity. In fact, the only thought running through my head right now is that, surely, if I had a boyfriend and a better social life, this would never have happened. Since when do I care about recycling and public transport? Or compost? Or blogs? I wish I'd never seen *An Inconvenient Truth.* You know what's really inconvenient? Thinking up 365 ways to green my life and writing about it every single day for an entire year.

What have I done?

This mixture of regret, confusion, bitterness, and pure embarrassment curdled in my stomach as I sat waiting for the replies to come in from friends and colleagues, all of whom I had just e-mailed about my challenge using as many self-deprecating adjectives as possible. See, whereas some people suffer post-Send-button anxiety after writing an overly effusive e-mail to an ex after a few too many glasses of wine, in my case it was simply because I had confessed something — and it was a confession that could very well jeopardize my nonchalant-pseudo-hipster cool status. I may as well have gotten down on my knees and said, "Forgive me, friends, for I have sinned: I've not only converted to environmentalism, I've started blogging about it, too."

Is it worse to be a blogger or a hippie? I don't even know.

In came the first response, from a friend who lives in Paris and works for the Associated Press. Matt eats food I can't pronounce, listens to obscure West African hip-hop, and basically attained cool status the day he was born; I'm pretty sure the only three R's he knows are Refined, Rioja, and Roquefort. Oddly, he's also a huge computer geek, and was on to me almost before I hit SEND.

According to an e-mail in my inbox, Matt had left a comment on my first post.

"Speechless, really," it said.

That was it, other than some quip about how the *foie gras* he was eating was most definitely "bio," which I think is what the French say when they mean natural, organic, environmentally friendly, or just hippie-approved in general, which is ironic because I'm pretty sure the last thing any hippie would be caught noshing on is *foie gras.* In fact, forcing a tube of lard down a bird's throat is more like a *faux pas.*

Then came an e-mail from my friend Jacob, who lives halfway across the world in Ramallah, Palestine, where he's starting up a nonprofit organization called Souktel, which uses text-messaging to connect employers with job hunters over their mobile phones. I wasn't sure what the specifics of it were beyond that—all I knew was that he put in fourteen-hour days in a conflict-plagued region, and most likely did not have time for blogs.

"Christ," began his e-mail, "you know, if it was anyone else, I would launch into my usual anti-blog tirade. But in this case, something tells me I'll soon end up adding you to my Twitter or Flickr or whatever, so that I can hang on your every enviro-word each time it comes into my inbox. I also vote for some live-blogging, or maybe vlogging, although I'm not quite sure what that entails."

Half-sarcastic, half-sincere. This was classic Jacob.

But within the hour, there came a wave of support, especially from my girlfriends. Most wrote to cheer me on, offer suggestions, and even reveal their own inner yearnings to stop wasting so much paper at the office or start bringing their lunches in Tupperware containers. One of my colleagues, Maryam, even wrote back, "I'll see your blog, and raise you a MySpace," boldly stepping out of the tech-nerd closet.

And yet, this didn't make it any easier. Now everyone knew: my family, my friends, my coworkers, and an already increasing number of environmentalists with enough time on their hands to putter around what my editors like to call "the online community," seeking out new green blogs. And who was this Lori V. anyway? She'd already left two comments on my site, and I knew nothing about her other than the fact that she also liked recycled paper towels. Either way, enough people were in on my challenge now that there was clearly no turning back—besides, no matter how ridiculous I may look with all this online journal–keeping and my amateur enviro-pursuits, in the end, I'll probably look even more foolish if I bail on it within the first twenty-four hours.

"Okay, just stop for a second," I told myself. "Here's what needs to happen: I need to take a deep breath, close my eyes, and just mentally dive into this green pool, headfirst. If I start drowning in compost, choking on all the crunchy granola, or otherwise find myself getting in way over my head, I can pull myself over to the edge, let the blog drown, and climb out soggy and smelly but otherwise unharmed."

Fortunately, I used to be a lifeguard.

Unfortunately, this was one deep pool.

But, as the Buddhists always say—and who can you trust, if not a Buddhist?—it's important to live in the present. So, I'm going to remind myself that today, all I'm doing is switching from regular paper towels to recycled paper towels.

Actually, they're 100 percent postconsumer recycled, unbleached paper towels produced with 80 percent less water than the industry average and dried with natural gas, the specifics of which I feel the need to relate because I'm already worried the imaginary Al Gore who's suddenly appeared on my shoulder will start reprimanding me for not abandoning disposable paper towels altogether and opting for reusable tea towels instead—or better yet, tea towels made from reclaimed wool that's been knit at a fair-trade establishment within one hundred miles of my apartment and shipped without packaging via bicycle.

But what can I say? I'm a soft-core environmentalist. When my cat, Sophie, decides to relieve her bowels on the floor instead of in her litter-box, there's no way in hell I'm picking it up with anything

I can't promptly toss in the garbage afterward. I thought about using those Bounty Select-a-Size ones, which let you tear away smaller pieces, but after further mulling, and a lot more spilling, I decided to resist the comfort of the brand and go for the unfamiliar but promising-sounding Cascades. The name Cascades made me think of waterfalls, rolling hills, and to a lesser extent a posh rehab clinic. I figured the beige color nicely complements my fake hardwood floors and while the towels may not be as strong or durable as the so-called quicker picker-uppers, I'm not exactly planning on soaking them in blue liquid and trying to carry large, heavy objects with them as the commercials always like to demonstrate.

So that's the first change done. On the pain scale of one to ten, I'd give it a two. Easy.

Now, I just have another 364 to go.

MARCH 3, DAY 3

Ban all Styrofoam

My friend Meghan, whom I met on the first day of high school and who lives just a few blocks west of me now, will probably turn out to be one of my greatest supporters in this challenge. This doesn't surprise me. We do a lot of geeky things together and then talk about how great said geeky things are. For example, we'll tie on matching aprons and cook organic, gluten-free soup and roasted vegetables at her place and trade halfsies for lunch the following week. We also download yoga podcasts and do the routines together. We rode our bikes over four hundred miles from Toronto to Montreal to raise money for a local charity helping people with AIDS. And of course, most recently, we both started up blogs—hers is called the Healthy Cookie and revolves around holistic nutrition; she used to be in advertising but then ongoing digestive problems led to a sudden change in career and now she's back at school learning about the enzyme properties of sprouted lentils.

The funny thing is, even when Meghan and I don't intentionally try to do things together, we usually end up together anyway; I'll be at a play and run into her, she'll be in the produce aisle at the supermarket and run into me, and so on. So it came as no surprise when, recently, I split up with my boyfriend and she became single, too.

At any given bar, we tend to be each other's good luck charms. Most likely, it has something to do with the fact that, physically, we're complete opposites—I'm tall and fair-skinned, she's petite and tanned; I have long, light brown hair parted on the side, she has short dark hair with a fringe; I drink red wine, she doesn't drink; I do the strong, silent bit, she works the cute, giggly thing. We cover all the demographics. There's something for everyone.

So when I found out about this anti-Styrofoam party—yes, an anti-Styrofoam party—I begged Meghan to come with me. It was being held by Get It to Go Green, an organization making life difficult for takeout restaurants as they try to convince municipal governments to ban polystyrene and replace it with something like the NaturoPack, which looks and feels the same but is made from corn, sugar cane, and potatoes and is completely biodegradable. Apparently, when they aren't fighting the evils of CFCs, these guys also like to get down. The party was held at a nearby hotel bar, popular with the West End indie set, so we figured there might be some cute boys in thrift store denim and ironic T-shirts who were into the green thing and would offer to buy us a round of locally brewed hemp beer or something.

Well, here's what we concluded within half an hour of arriving: within the urban hipster community, there is really a whole subdemographic, best described as the eco-hipster. These guys look more or less the same as any other ironic-sunglasses-wearing, Broken Social Scene listening hipster but are in fact an entirely different breed. In place of that espresso and cigarette smell, they're more redolent of beet juice and pot, and unfortunately, they often lack the cynicism chromosome. This can sometimes be endearing, especially if they're trying to pedal a stationary bike that's hooked up to a generator powering a string of LED Christmas lights on the ceiling, or cheering on an environmentally aware rap group without the faintest glimmer of embarrassment, but they have their fair share of problems, too.

See, the eco-hipster set can subsequently be divided even further, into the eager beavers and the serious activists. If you were to, say, make a joke about the vegan community, the serious activist would be offended, whereas the eager beaver simply wouldn't get it. While neither would be caught dead sitting in front of a television set on a Wednesday night rating the fierceness of Tyra Banks and her top-models-in-training, the serious activist's reason is because he has more important things to do; the eager beaver just thinks mean people suck and has a badge sewn on his backpack to prove it. The activists in particular, though, seem to lack not just a sense of cynicism but a sense of humor altogether, and while their intense gaze can draw you in, it's usually the case that they're less intent on getting your phone number than they are on debating the efficacy of bio-fuels. Not much of a turn-on.

But Meghan and I tried. We signed their petition, ordered an organic beer and a glass of water, and attempted to zero in on a few of the cutest boys in the crowd.

"What do you think of him?" I asked, pointing my bottle in the direction of a guy who looked as though he might, at one point, have been offered a walk-on role in a romantic comedy starring Matthew McConaughey, Jennifer Aniston, and a manic but ultimately endearing puppy, then at the last minute decided acting was too shallow a career path and followed the green movement instead.

"Meh," replied Meghan. "I'm not so into that sweater. It makes me think of Bill Cosby. What about that guy over there?"

I looked to where she was pointing.

"Kind of potato-faced," I said.

This is typical of our conversations about men. Obviously, we're both hypercritical, to the point where we're dismissing some poor guy over such a trivial thing as the mosaic pattern on his sweater or the starchiness of his visage. But first impressions do count, and we never argue the other's verdict. Besides, it took only about twenty minutes before we'd settled on some mutually approved prospects.

As a girl took the stage wearing a HOTTER THAN I SHOULD BE T-shirt—a climate change joke—and muttered something about why we were all there, I leaned in and asked my guy, with as much eyelash batting as possible, if he'd heard what the host just said because I could barely hear over all the bike-pedaling.

"She's introducing the first act," he deadpanned, and faced the front again. I shuffled closer, but as soon as the group launched into their enviro-rap, which sounded like something a kindergarten class might perform for their teacher as an Earth Day project, he suddenly got really into it, jumping around, pumping his fist in the air, and hollering encouraging remarks at the stage. A serious activist. With no rhythm. And even less self-awareness.

Forget that.

I turned around to see how Meghan was doing with her boy, the ticket collector and hand stamper at the door. Within seconds, she was back at my side.

"Brutal," she said. "I asked him if he knew when the headlining act would be playing because their CD was sitting on the table and it looked cool."

"And?" I said.

"And he told me, 'Oh yeah, they're on next—but you could just buy the CD and go home to listen to it.' I mean, he basically told me to leave!"

Maybe we were overanalyzing things and not being aggressive enough; perhaps our expectations from two hours at an anti-Styrofoam party were too high. Either way, we decided to take the ticket collector's advice and call it a night.

MARCH 5, DAY 5

Turn down my thermostat

Farch, as its name denotes, is not a pretty time of year. It's the general season encompassing February and March when it's no longer winter but not quite spring. There's no blanket of freshly fallen snow, nor are the birds chirping or the trees budding; instead, there's just gray skies, gray streets, gray slush, and in my particular case, gray moods. So I thought maybe if I took some initiative by shoving my salt-stained boots and snot-encrusted scarf in the back of the closet, buying a fresh bouquet of tulips and turning down my thermostat, Nature would get the hint and follow suit with some sun and warm weather.

This last part especially, switching off my heat, seemed like it would make for another easy green change. Granted, I may need a few more layers and some extra cups of hot chocolate on particularly dismal nights, but it wouldn't be for long. Besides, I was already eating salads for lunch and letting my old friend Visa treat me to a new pair of sunglasses—magical thinking or not, I could feel spring right there.

Sometimes, however, I forget that my circulation has been on strike since '79 and the union that is my blood pressure and heart rate is not willing to compromise. My fingers and toes are rarely any color besides red, purple, or sometimes white—even in the summer, if I walk into an air-conditioned room, they show signs of frostbite. Within days, then, I found myself going to bed with three pairs of wool socks, mismatched flannel pajamas, gloves, two extra blankets, and a hot-water bottle tucked into my pillow, all the while hoping that Sophie might sit on top of me for yet another layer of warmth. During one especially frigid night, I switched my alarm-clock radio over to a station that was playing some calming music, hoping to soothe my hypothermic self to sleep.

The next morning, as I woke up to the easy-listening sounds of Kenny G, spooning the pillow with my hot-water bottle, onto which my nose and salivary glands had evidently been leaking all night, it occurred to me that none of this would have happened if a) I'd left my thermostat alone, b) I lived in one of the hundreds of perfectly nice cities below the 49th parallel, or c) I had a boyfriend in bed with me to heat things up.

MARCH 6, DAY 6

No more bottled water

My favorite stand-up comedian, Jim Gaffigan, has a line in his routine about the bottled water phenomenon. It comes somewhere in between his bit about the redundancy of Mexican food and the appeal of the manatee. He asks, "How did we ever get to the point where we're paying for bottled water? When they first brought it out, I thought it was so stupid . . . Then I tried it and thought, 'Yeah, this is good. This is more watery than water. This has a water kick to it.'"

Somehow, this laughable prospect of paying for water became a reality, and I'll admit to being sucked into it. I've become a bit of a water snob, too. I mean, I've always been a bit of a snob about most things, but when it comes to bottled water, I know my Evian from my Volvic, my San Pellegrino from my San Benedetto, and I'm not about to spend good money on tap water that's gone through reverse osmosis and had three kinds of salt added to it by the fine people at Coca-Cola when I could be drinking natural spring water purified by equatorial winds and bottled at the source of a primitive rainforest in Fiji. On the other hand, I am old enough to remember when bottled water first hit the market and became a craze, and part of me knows that the entire concept of bottled water—at least in a developed country—is, as Gaffigan points out, completely ridiculous, not to mention a huge waste of plastic and unnecessary shipping.

So for my next green change I'm going back to drinking plain old tap water. I have a Brita filter, which means at least it won't taste so metallic, and I can carry around a Nalgene bottle if I'm going to be out all day. Although this accessory is hardly cool—in fact,

I'm pretty sure most stores have a policy against bringing any dogs, outside food, or lame water bottles inside—it does bring back memories of being at camp and twisting my hair into symmetrical braids, paddling a dilapidated canoe out on Big Hawk Lake, and working on my Teva tan as I slapped the occasional mosquito, sang Kenny Rogers tunes (out of tune, admittedly), and sipped warm water from my sixteen-ounce loop-top Nalgene.

Recently, I've been reading all these articles about how certain polycarbonate plastic bottles have been shown to leach bisphenol-A, an estrogenic hormone, as well as create a breeding ground for nasty bacteria if not properly washed. But I figure, I clean mine every day with dish soap, and what's a little top-up of estrogen really going to do anyway? I might have less chance of getting knocked up and fewer zits.

In my world, that's called a bonus.

MARCH 9, DAY 9

Eat locally (within the United States and Canada)

I've started to face some problems in the grocery store, particularly in the produce section. To begin with, I usually find myself standing, basket in hand, by the rows of apples, unable to choose between an organic one from the United States or a pesticide-sprayed one grown locally here in Ontario. The organic American ones don't look as shiny, but maybe apples aren't supposed to be that shiny; I could always clean the pesticide off the local ones at home with a special fruit and vegetable wash, but that just requires more product consumption. I often turn to ask that mini Al Gore who popped up on my shoulder when I switched to recycled paper towels, but my imaginary friend can be pretty flaky when it comes to making tough decisions.

If Meghan accompanied me on these grocery trips, I know she'd say that my health is more important than the planet's and that organic produce has been shown in numerous studies to have more nutrients, so I should just support the farmers who are trying to do things the natural way and get the American one. But I also know that humans have been relying on their senses for centuries to gauge what's good for them and what's not—which is why the sight and smell of rotting food sends yuck signals to our brains—and those Ontario Royal Galas usually look pretty darn good. They're also guaranteed to be fresher, which means they're probably better tasting, and because they didn't require days on the highway bumping about in an eighteen-wheeler, I'm willing to bet they're just as nutritious. Still, does buying the local apples mean I'm creating a demand for and thus encouraging the use of pesticides?

The last time I encountered this dilemma, I stood around, staring at the apples for at least five minutes, which doesn't seem that long but in grocery-shopping time almost demands a call to security over the intercom. Of course, it would be perfect if I could eat food that's both local and organic, but it seems that just about the only people able to do so are Californians—other climates are either too wet, too dry, too hot, too cold, too dark, or too sunny. For the rest of us, there will always be that little bit of stress here in the produce aisle. As the best-selling food author Michael Pollan has pointed out numerous times, our collective knowledge about what to eat has fallen by the wayside, leaving us with food pyramids, dieticians, fat camps, and entire books devoted to deconstructing the Twinkie—a treat, by the way, that contains over ninety ingredients, most of which are man-made, one of which includes beef extract.

I'm convinced, ecologically speaking, that buying local is more important than buying organic, but I'm not about to restrict myself to the borders of Ontario—at least not yet, unless my doctor prescribes me a locally procured antidote to scurvy. I will, however, only buy food that comes from within Canada and the United States. This means, right off the bat, that mangoes, avocados, and bananas are outlawed until the end of the challenge. Lemons and limes are also out, but only until summer, when California grabs the citrus reins back from Mexico.

Alcohol, however, will remain unrestricted, at least until I find a local wine that doesn't taste like Welch's grape juice.

MARCH 14, DAY 14

Switch to phosphate-free dish detergent

I don't think there's anything more toxic-looking in the cleaning supplies aisle than those tiny plastic pouches of neon goo and white powder, otherwise known as Cascade 2-in-1 ActionPacs. But the fact is, when it comes to dishwashing detergent, they really get the job done, and who doesn't love that steaming hot, bleachy-clean smell that wafts up when you open the dishwasher? My favorite kind is the one that comes with extra bleach and a "Fresh Rapids" scent. While I'm not exactly sure what rapids smell like—let alone how fresh rapids might differ from stale rapids—I do like the idea of turning on my dishwasher, hearing that familiar buzz, and picturing an enormous wave rolling about inside, crashing gently but effectively against all my plates and bowls until they're invigorated, gleaming and gunk-free.

Sadly, the other day, I popped my last pouch and had to go looking for a safer, more natural alternative.

"Why don't you try using baking soda and vinegar?" asked the imaginary Gore, who now chose to make an appearance as I stood in the aisle at Whole Foods. Can I fire an imaginary friend? Either way, he started to explain the benefits of simple, homemade detergents and disinfectants, and I did actually consider it, but baking soda and vinegar always make me think of those elementary school science experiments, where things froth up and start to smell funky, which is not what I want swishing around my high-end stainless steel dishwasher. Ultimately, I decided to try out this brand called Seventh Generation instead.

Based in Vermont, they offer one of the most widely available natural product lines around, a bit like the financially secure but pot-smoking cousin of Procter & Gamble. Their website features press releases carefully positioned next to pictures of mothers holding babies, children holding puppies, and seniors holding watering cans. You can follow links to scientific data on all of their natural ingredients or, far more entertaining, to a blog called the Inspired Protagonist, with the mission statement "Cutting the ties of negativity that bind us." Oh, Vermont.

Well, I tried the detergent once, then tried it again with double the amount, but it still cleaned only about 80 percent of the dishes, the spoons faring the worst. And when I opened the dishwasher door to let the clean smell waft up, there was more of a neutral-to-somewhat-stale-cheese odor.

I want my phosphates back.

MARCH 16, DAY 16

Use tote bags

Say the words *tote bag* and I immediately think of a dirty canvas sack, the color of seaweed and expired cream, sitting stiff and crunched-up in the bottom of my parents' linen closet. My mother got it for free at some medical conference and, when I was a kid, we used it only on the rare occasion when the entire family was going to the local pool and needed something in which to throw our wet, chlorinated swimsuits afterward. It has some generic logo and acronym screened onto it that no person would ever recognize, unless that person was a radiologist. In 1984. In Orlando, Florida.

There is truly no fashion accessory more maligned than the canvas tote, but I needed something to bring to the grocery store so I didn't have to use as many plastic bags. My backpacking years—along with my Kraft Dinner–eating, essay-writing, and hosteling years—are officially behind me, so a knapsack wasn't an option, and those wire pushcarts are an even worse aesthetic crime; if I could just find a not-so-heinously-colored tote bag, though, perhaps one that came in a more lightweight material than canvas, my problem would be solved.

Finally, I found it: a collapsible nylon tote, plain white with an adorable flower print on it from a company called Three Wishes.

It folds up into the size of my fist, so I can tuck it into my purse, and it expands to the size of about two regular plastic bags. When I bought it, I thought this could very well turn out to be the best green change of the month. Having it on me at all times would mean that even if I made an impulse purchase—which, let's be honest here, happens almost daily—there's never an excuse because I have a bag already.

I decided to give it a test run.

Actually, I decided to go shopping for some new clothes and justified that as a way to give my new tote bag a test run. Walking by one of my regular stores, a brand-name chain that specializes in sweatpants and purses, I saw a sign in the window advertising a sale where everything green—as in the color—was half-price. Somehow this was meant to spread environmental awareness, though it didn't appear any profits were on a direct path to Greenpeace. Anyway, I went in, tried on a shirt, decided to buy it, and declined a bag. I explained with a proud smile that I'd brought my own, then unfurled the Three Wishes tote with a magician's flourish and waited for the cashier to let out some twitter of encouragement.

Instead, he just convulsed slightly and cocked his head.

"Why-y?" he asked, giving the word an extra syllable to denote confusion and possibly a mild form of disgust.

"Because I have my own, right here," I said, adding that I thought I was complementing the green theme of the day by reusing instead of consuming.

"Uh, okay, I guess so . . . if you really want," he said.

What kind of a response was that—if I really want? Why would I want otherwise? For the sake of walking around the block with the logo of a semioverpriced retailer under my arm? For the convenience of throwing it out when I got back home rather than rolling it up and storing it away?

I think I'm starting to understand why the hardcore tree-huggers get a reputation for being so defensive and huffy.

MARCH 18, DAY 18

Cancel cable

Don't get me wrong—I do want my MTV, I just don't want to pay $50 a month for it. As much as I've grown attached to all of the Bachelors, Bachelorettes, and Survivors; Donald Trump and his wannabe entrepreneurs; and the brilliantly effusive hilarity that Colin and Justin bring to the reality show *How Not to Decorate,* I've also realized that my bank account is not such a big fan of all this. Looking at my monthly statements, I knew something major needed to be cut. Internet was essential, as was heating, electricity, and water; my job meant I had to keep my cell phone but I'd already gotten rid of my landline; I was trying to purchase fewer clothes and be more strategic at the grocery store, too. But still, with one foot in the black, one in the red, and an uncertainty of where all this green stuff would lead me financially, a savings of $50 a month was pretty significant.

So I called the local cable company to cancel my account.

"Oh my god, I'm so sorry to hear that," said the operator, when I told her the news. Hearing her tone of voice, one might have guessed I'd just told her my cat died.

"Don't worry about it," I replied. "It happens."

"May I ask why?" she said.

"Yeah, sure. I mean, frankly, I can't really afford it."

"Oh, I see . . ." she said, solemnly. "Well, what if you changed over to a reduced-rate plan? Basic cable would still give you all the lower channels—"

"But it wouldn't give me Tyra now, would it? Seriously, just go ahead. Cancel everything. I can't even talk about it anymore."

And with that, the deed was done.

Sitting there a few minutes later, my pink cell phone still in hand and warm from the conversation, I couldn't help but feel that something greater should come from this, something more than just money in the bank. Now that I'm in the routine of writing about all the changes I make to my life, it seems like I should be blogging about this. But canceling my cable wasn't green in any way.

Or was it? It would mean I'd be using less electricity to power a television set for three to four hours every night, not to mention the battery power required for the remote control. Granted, this doesn't add up to thousands of kilowatt-hours, and it's something I would have done regardless of my challenge, but even if it makes only a small dent in my carbon footprint, why shouldn't it count?

Backwards logic, yes. But just because a green change falls into my lap doesn't mean I'm going to ignore it. It may still be early on, but I'll take what I can get.

MARCH 19, DAY 19

Switch to corn-based, biodegradable cat litter

My cat, Sophie, has three favorite activities: 1) licking yogurt from my spoon when I'm not looking; 2) meowing to go outside on the balcony and then, after all three sliding doors are pulled back, sniffing and walking away; and 3) defecating on my bed. She's a British blue, a breed that nearly became extinct in the sixteenth century as the more superstitious folk in Europe began to suspect they belonged to witches and killed as many as they could. Although I don't believe in witches, I am confident that Sophie is, in fact, a descendant of Lucifer—and if not the devil himself, then at least that nasty cat in *Cinderella.* I've spent the past eight years buying her different food, treats, toys, grooming brushes, everything and anything that might make her like me enough to purr every now and then, or at the very least stop crapping on my bed.

A few years ago I tried changing her litter from a clay-based variety to a silica-based one, but that didn't seem to do much. The silica stank less, but tracked pebbles all over the floor, and whenever I stepped on one by accident it felt like I'd rammed my heel into a thumbtack—or actually worse, a thumbtack covered in week-old cat urine. So when I saw a new corn-based, biodegradable litter at the pet store, I figured it would at least be worth a shot, if only for the sake of this challenge. Plus, it was called World's Best Cat Litter, so it couldn't be that bad.

In fact, the name turned out to be more than justified—this stuff smells nice, it doesn't track and, most importantly, Sophie loves it. She hasn't once left a parcel of evil on my bed since I've made the switch, and this afternoon, I think I even heard her purr (although she might have just been preparing another hairball).

MARCH 24, DAY 24

Sign up with local Freecycle network

I'm a horrible product junkie, a marketer's dream. If it comes in a little white jar and has a word like *radiance, glowing,* or *firming* on it, I'll buy it. Because of this, my bedtime ritual has, over the years, gone from the most basic ablutions—brush teeth, wash face—to an eighteen-step regimen, which, strictly above the shoulders, goes something like this: brush hair, brush teeth, brush tongue, floss, use mouthwash, remove makeup, wash face, exfoliate, apply toner, apply wrinkle cream, apply freckle-fading cream, apply eye cream, and, finally, apply zit cream, if needed. Even worse is that because I work in the arts section of the newspaper, companies send me a lot of swag—free samples of their newest lotions and potions that I often use and never write about. But I'm trying to declutter my bathroom counter and invest in a select few natural products that don't have any potential carcinogens or other toxins. This means that I need to part with all these jars and bottles of stuff, most of which I've tried only once anyway.

I don't want to throw it all in the garbage, but none of my friends really want anything I've dipped my finger into, and even if it's something I haven't opened, most girls I know already have their own product loyalties. I'm not sure I can drop something like this off at Goodwill—in fact, I can't even take most of it on a plane anymore—but then I read about this online network called

Freecycle, which is a bit like Craigslist except there's no money involved; it's just a group of people living in the same city offering stuff and taking stuff for free. So I signed up.

Immediately, the e-mails began pouring in from my local group. I noticed they all had subject lines like, "OFFER: Gardening tools," or "WANTED: Children's books," or if someone had already snapped up an item, there'd be one saying, "TAKEN: Sewing machine." When I opened each e-mail, there was usually a more lengthy description of what was being given away and why, as well as where the person was located. For the most part, it was all junk. Take, for instance, this post: "OFFER: Couch"—sounds great, a new couch. But then you read further and it says, "Brown velour couch, had it for 20 years, only a few stains!" Gross. Or there's the stuff like, "OFFER: Three floppy disc boxes"—does anyone still use floppy discs? And if so, don't they already have the boxes they came in? My favorite posts, though, were the ultraspecific wanted ads like, "WANTED: Five old clawfoot bathtubs" with the esoteric explanation "I'm doing a competition and need these soon!"

I posted my own note that read "OFFER: Deluxe toiletries" with an additional note that I'd gotten a bunch of wrinkle creams and body lotions at work that I wasn't going to use and was looking to give them away; then I gave the closest intersection to where I lived and said it was available for pickup only. Within thirty seconds someone named Buddy replied. He happened to live a block away and was looking for a last-minute gift for his wife's birthday. Why his wife would want to receive a bunch of used wrinkle cream for her birthday I have no idea, but I wasn't about to ask questions—I just wanted to get rid of this stuff. So Buddy and I spoke over the phone and arranged for him to drop by the next morning.

At the appropriate time, he buzzed my apartment. I was of course still asleep, it being the weekend and all, but quickly rolled out of bed, gathered all the stuff together in a gift bag from under my cupboard (I couldn't hand him such fancy products in a plastic grocery bag, surely). Then I schlepped my way down to the front of the building in a pajama top and sweatpants. As I walked down the hallway, I could see Buddy standing outside. He had to be at least fifty, had a fuzzy beard, and was wearing a camping vest and Tilley hat. It was drizzling, but he didn't seem to mind. I opened the door, a little cautiously, and stepped out.

"Um, Buddy?" I said.

"Hi, that's me!" he replied and reached out to shake my hand. I couldn't believe I was standing outside my building in pajamas, handing over a bag of assorted wrinkle creams to a man my father's age in a Tilley hat. I wasn't sure what the Freecycle protocol was, either. Do I invite him in for coffee? Ask if there's anything else I can do for him? Give him my card? Thankfully, Buddy came to the rescue and simply made a nice comment about how happy his wife would be to get this, wished me a pleasant weekend, and was off. He even sent a thank-you e-mail the next day.

I have to admit, as Buddy walked away and I headed back upstairs, I felt a quick altruistic rush. It's not as though Buddy or his wife were in desperate need of wrinkle creams, and I'm sure I would have found some other way to get rid of them rather than throw them in the garbage, but I liked being able to take part in this grassroots, community-based initiative and meet someone from my neighborhood in the process.

MARCH 30, DAY 30

No car on the weekends

I'm only a couple of weeks into this challenge, and already the competitive side of me is surfacing. I'll be scanning the aisles of a health food store looking for fair-trade, GMO-free, organic whatnot and then a woman will walk up beside me and tuck a bag of quinoa into her stylish, logo-free tote bag. Whether intimidated by the phonetic mystery that is "quinoa" (pronounced *keen*-wah, I now realize) or the fact that her bag is prettier than mine and probably made out of repurposed hemp instead of nylon, I'll suddenly feel like I'm in an all-out green-off and frantically start looking for ways to one-up her. A fluoride-free baking soda and tea-tree oil toothpaste with recyclable packaging? Organic raw almonds harvested by ridiculously well-paid workers and transported by monarch butterflies during their annual northern migration?

I'm green with envy of another person's green status, and it's pathetic. I've been feeling the same way whenever I stumble upon another eco-blog—like No Impact Man, for example, who's basically doing what I'm doing and taking it to a whole new level, living completely impact-free with his whole family in downtown New York and not using toilet paper or taking elevators—and so I worry that if I can't be the greenest, I should just quit, throw my recyclables in the garbage, and leave the tap water running. But this is the wrong mentality, of course. We should all be in this together—not just environmentalists but everyone, from No Impact Man and Al Gore to me and all the other green-hearted but skeptical people out there, because as the Discovery Channel reminds us daily, we've only got one planet.

Arguably even worse than green envy, though, is this other feeling I've begun to develop: the need for approval from the greenkeepers—that is, the editors at big eco-blogs like TreeHugger.com, the owners of my local grocery co-op, the organizers behind that anti-Styrofoam party, and so on. Even if I've managed to suppress all my insecurities about whether I'm the most environmentally conscious shopper in the health food store, by the time I get to the checkout and proudly answer the question "Paper or plastic?" with "Neither, thanks, I brought my own," I'm practically shocked if the cashier doesn't smile approvingly and give me a gold star.

This is a problem. This is why hippies become smug. It's so easy to start expecting a reward greater than simply knowing you've done something good. Even if you walk out of the store into a park and inhale the fresh air, it's too late to make an immediate connection between cause and effect, between a cluster of raw organic almonds and a cluster of trees still standing. And maybe there isn't such a direct link anyway—maybe purchasing the right environmentally friendly products doesn't do anything other than force the multimillion-dollar companies to step up their marketing with the appropriate "natural" and "organic" labels in yet more green competition and, inevitably, empty promises.

I don't know when I'm supposed to recognize that the world is officially better because of my little green steps, but the point is that there's some kind of place between irritating, go-getter environmentalist and jaded, smug hippie that I need to find.

It helped today that spring finally graced those of us in the Great White North with her presence. Watching as my surroundings turn greener by the day makes me feel as though my changes are adding up. Maybe the grass in the park across the street looks healthier this year because I haven't been using chemical-based products. Maybe the streets look cleaner because I haven't bought anything packaged in three layers of plastic recently. Or maybe it's just that I'm more in touch with Nature now that I'm jogging outside instead of on an electric treadmill.

On the one hand, it's hard to believe a month has gone by, but on the other hand, I can't believe I have eleven more to go. I don't look disheveled or smell like patchouli, and my apartment appears pretty much the same except for some different products here and there, but I have a suspicion that it won't be long before I'll start running out of ideas and be forced to make some big moves. Already, I'm having moments when I dread having to think up yet another green change, when the mere word *green* makes me cringe. At work the other week, I had to sit through an epic, thirteen-hour French film by Jacques Rivette, which was unbearable at first, but eventually I began to appreciate its tortured, Euro aesthetic. When I finally walked out of the theater and into my own reality again, switching from philosophies about Molière and existentialism to whether I should start composting or switch to an organic shampoo, it all seemed so stupid. Only a privileged North American with too much time on her hands would spend an entire year of her life worrying about whether her deodorant had aluminum in it or how much it would cost to carbon-offset a flight to Montreal. On my way back home that night, I tried to think of a way out—something I could tell my boss, my family, my friends, my readers, that would let me off the hook without too much guilt. But when I sat down at my computer and pulled up the website, it brought up a statistics graph. I stared at the little meter that had counted over one thousand hits that day. I stared and I stared.

That number represented one thousand pairs of eyes—eyes that belonged to people from as close as Meghan's apartment to as far away as Australia. They were eyes that looked at what I was doing and read what I had to say about it. Maybe nothing would come of it and they'd just move on to PerezHilton.com for the latest news on J.Lo or Brangelina, but maybe a few people would actually connect with my struggle to buy the right apple at the grocery store, my assessment of the eco-hipster demographic, or my somewhat bizarre Freecycle experience. Perhaps the occasional blog post would inspire a reader to go out and buy some recycled paper towels or a not-too-ugly tote bag.

But even if it doesn't—even if the only thing this self-imposed challenge does is get me to reevaluate my shopping habits, offer a few sacrificial light bulbs at the feet of Mother Nature, and figure out what it truly means to be a modern environmentalist, then ultimately, it's worth it. That cheesy saying—about always regretting what we haven't done, never what we have—is what I'll have to cling to in the months to come.

The only thing holding me back right now is fear (isn't it always?). It's day 31 and already I'm running out of ideas. This goal of making baby steps could very well prove impossible, simply because there are only so many product switches I can make and only so many cat-related changes I can thrust upon poor Sophie. As well, if I approach this in the same way I approach any homework assignment, my perfectionist and obsessive-compulsive tendencies are sure to kick in, which could very well translate into a neurotic, self-destructive enviro-addiction. By the halfway point, I'll probably have gone beyond the endearing crunchy granola thing and be living off the grid as a practicing Luddite, eating nothing but sprouted lentils grown locally in my own compost.

This, then, is the dilemma at one-twelfth of the way through:

Can I cradle a martini with dirt under my fingernails?

Seriously. I mean, is it really possible to make hundreds of lifestyle changes without becoming someone I'm not? How do I sink myself deep into this green movement without losing perspective? Actually, forget perspective—how do I not lose my job? Or my apartment? Or my sanity?

The answer, I think, is to tether myself to all the rational people I know, and just pray they don't fall in with me.

april

1	Use a thermos for coffee and tea
2	Drink only fair-trade, organic, locally roasted coffee
3	No more gift-wrap unless it's used or scrap paper
4	Switch to vegan-friendly dental floss
5	Change all light bulbs to CFLs
6	Use beeswax or soy-based candles — no paraffin
7	Recycle all beer and wine bottles
8	Switch to natural toothpaste
9	Use a natural laundry detergent
10	Eat only free-range, organic, hormone-free (and if possible, local) meat, restrict to once per week
11	No more petroleum-based loofahs and bath poufs
12	Sign up with GreenDimes.com to block junk mail
13	Keep track of water usage — use a trickle of water for handwashing and brushing teeth and limit showers to five minutes
14	Use the air-dry function on the dishwasher
15	Consume only locally brewed beer (organic when possible)
16	No more paper towels or hand-dryers in public bathrooms
17	Invest in permanent laser hair removal rather than shaving and waxing
18	Only local and fair-trade chocolate
19	Turn off my freezer
20	Pick up litter whenever I see it
21	Dispose of all used batteries at a local hazardous waste depot
22	Switch to an eco-friendly dish detergent
23	Use natural, paraben-free lip balm
24	Consume only wine that is locally grown, produced, and cellared
25	Incorporate baking soda in household cleaning
26	Switch to natural shaving cream
27	Switch to recycled, recyclable razors
28	Use only one glass or one mug per day
29	Spend part of each day learning about environmental issues
30	No more nonstick frying pans

APRIL 3, DAY 34

No more gift-wrap unless it's used or scrap paper

My sister and I have the same voice, the same hair, and the same parents, but that's pretty much where the similarities end. Five and a half years younger than me, Emma tends to float through life, going wherever the wind takes her. I mean this quite literally—if there's a strong enough gust, chances are that her ninety-five-pound frame will be picked up and blown halfway across the city, which will actually be fine with her, as long as there's a Starbucks nearby. It's not as though Emma's without her anxieties, of course, but she's good at either forgetting or suppressing them because, on the surface, this girl is nothing if not the model of placidity. Much like Bambi's, all of my sister's physical features are tiny except for her enormous brown eyes, which means she gives off this look of fey naiveté, which is for the most part fairly accurate, although Emma has had just as much exposure to the Britishisms of our parents while growing up and therefore has managed to carve out a dry, caustic edge in her personality, too.

There has been some sibling rivalry in the past—almost entirely initiated by me—but we get along now, especially if we talk about clothes, boyfriends, pop culture, and family politics. When it comes to all my environmental do-gooding, though, there's not an inch of common ground between us. Emma isn't exactly out to harm the planet, but she doesn't seem to care much about helping it, either. She'll put stuff in the recycling bins at home because they're there, but she's very much part of a generation that places a high importance on brands and logos. Once, when she got a summer job at a downtown clothing store, Emma saved every penny from two months' worth of paychecks and, on her last day, spent all of it on a Louis Vuitton handbag that was about the size of the croissant she ate that morning for breakfast. She also has the disturbing ability to, in a cursory glance at the back pocket of a pair of jeans, immediately determine whether they're Sevens, Fidelity, or Rock & Republic and subsequently will explain what that implies about the person wearing them. The general rule is, the more expensive the design, the cooler you are. (Although Emma believes the key is in the subtlety; any loud proclamation of money is tasteless.)

This evening, while out for dinner and drinks with my sister, I was trying to explain why I'd decided not to use any gift-wrap for presents and instead opt for old newspaper, reusable fabric, or nothing. Needless to say, Emma didn't get it.

"But that's, like, part of the present," she said. "And how does giving someone an unwrapped present stop all the icicles from melting up north, anyway?"

"You mean the ice caps? In the Arctic?"

"I thought an ice cap was short for iced cappuccino," she deadpanned.

"Har, har," I said. "No, I'm talking about the melting ice caps, about global warming, about how it's important not to waste paper and cut down trees."

"But it's not a waste, it's pretty. If I see one of those baby-blue Tiffany bags under the tree at Christmas, that makes me almost as happy as whatever's in there."

It's true. Emma derives a great deal of aesthetic pleasure from gift-wrap, packaging, and shopping bags. She's even been known to walk around the same block twice if she has a Tiffany bag in her hand, for the sole purpose of showing it off.

"You know," she said, "if there was a tote bag that looked exactly like a Tiffany bag and never got crumpled or dirty, I'd use that instead of disposable ones."

I rolled my eyes.

I had to admit, though, it wasn't a bad idea.

APRIL 4, DAY 35

Switch to vegan-friendly dental floss

There's a lot of weird stuff on the shelves at health food stores: Nayonaise (mayonnaise without the eggs), carob chips (dairy-free chocolate chips), Neatballs (soy-based versions of meatballs). But as I was browsing my way through the hygiene aisle the other day at Noah's, I saw something even more bizarre sitting alongside the baking soda toothpaste and myrrh-based mouthwash: vegan waxed dental floss. That's exactly what the label said, too: vegan waxed. Not vegan friendly, or naturally waxed, but vegan waxed. Reading the package, which was made of cardboard and therefore recyclable, it became somewhat clearer: the company, Eco-Dent, claims there's no beeswax or petro-waxes involved in the manufacturing, and the thread is treated with fourteen essential oils. It also includes one hundred yards of floss in the package, which is about three times as economical as most other brands.

It's a bit of a stretch, though; I mean, what brand of floss *does* use beeswax? And if it did, would vegans even care? Are they that strict about their dental hygiene? I decided to find out.

The only problem was, I didn't know any vegans.

So I walked around the corner to a vegan restaurant, ordered a large salad with tofu cubes, and asked the wait staff.

Fresh, a popular franchise in Toronto operated by a staff of Gen-Y kids with lithe arms, pale faces, and necks wrapped in either a bandana, keffiyah, or crocheted scarf, specializes in smoothies and vegan food. Despite its stereotypically vegan-looking servers, the restaurant's menu is actually quite varied—yes, there's tofu, tempeh, brown rice, and royal jelly floating around, but there are also things like sweet potato fries with miso gravy, Thai veggie burgers with side salads, split-pea soup with corn bread, carrot cake, brownies, and even booze.

While waiting for my salad, I asked the hostess if she'd heard about vegan waxed floss. I thought she might have some opinion on it, considering she's clearly at a higher-than-average risk of getting spinach and broccoli caught in her teeth.

"Um, no," she said. "I've never really thought about floss being an issue."

"Do you care about bees and their—um, you know—excretions being used for modern conveniences like, say, cavity prevention?" I asked.

She replied that she didn't. She wasn't a "hardcore vegan."

"It really depends on the person," she added. "It's like, you've got all the vegetarians, but there are also pescatarians and flexitarians—there are so many eating regimes out there now and people have totally different concepts of what's right and wrong. In fact, my manager says she's strict vegan, but I've seen her wear leather boots."

What?! No way.

"Yep," she said. "I guess they could have been secondhand, but they looked like new."

I asked her if she thought the vegan movement was more about animal welfare or environmental ethics, or if they were intrinsically related, and she said it was usually a little bit of column A, a little bit of column B.

"But then there's the odd person who just doesn't like meat or eggs and is allergic to dairy, so they kind of just become vegan by default."

Vegan by default. Somehow, that sounded like such a postmodern predicament.

Either way, I decided to buy the vegan waxed floss, if only for the fact that I could now justify eating more honey.

APRIL 5, DAY 36

Change all light bulbs to CFLs

I mentioned before that I'm a water snob. Well, as it turns out, snobbery extends into various aspects of my life and another one of them is lighting. There's no point in pretending—this is completely tied to my vanity and insecurity over freckles, wrinkles, pimples, chapped lips, and any other sort of imperfection that might be rendered glaringly hideous in the wrong light. So it was with trepidation and a sort of pre-regret that I decided to try compact fluorescent bulbs. It got worse at the store, where I was confronted with row upon endless row of spirally bulbs; and of course I'd neglected to check the wattage beforehand. I ended up standing there like a deer in energy-saving headlights, mouth probably open, unable to determine which brand was the greenest of them all.

In the end, because I had four bulbs to change, I decided to buy two of the NOMA soft white and two of the General Electric *plus compacte!* (the label appeared *en français* but *pas en anglais* for some reason), both of which were 13 watts, equal to 60 watts, which sounded about right. The advantage of the former was that the light would presumably be more romantic; the latter, on the other hand, were smaller—and thus cuter. Both, however, came sealed in that blister-pack stuff that isn't recyclable but is guaranteed to drive you mad and actually give you blisters while trying to open it until you give up, go on a hunt for scissors, and finally start jabbing at it in a murderous rage. Plus, the bulbs themselves contain mercury, which means throwing them out in the regular trash isn't an option, unless you're okay with poisoning rivers, streams, and the water supply.

Nonetheless, I needed to make this change if only for the sake of saying I tried them, and so I returned home with my new spirals of fluorescence, screwed them into the lamps on either side of my bed, as well as my desk lamp and reading lamp downstairs, then flicked the switch.

Nothing happened.

Wait, no, something was happening.

There's a split-second delay, it seems, before these things decide to come to life, blinking tentatively a few times, as though they actually realize how dismal they're about to make everything look and therefore feel obliged to warn those within an eight-foot radius to evacuate the area while they still can.

In fact, I shouldn't even use the term *come to life* because these bulbs emit more of a cold, morguelike glare. Even with lampshades, they're no match for my beloved incandescents. Granted, the light isn't quite as nauseating as that given off by the giant fluorescent tubes looming over my cubicle at work, but how could I possibly relax when my retinas were being so heartlessly assaulted by the brutal aesthetic reality of energy savings? Even if I did manage to find a boyfriend in the midst of all my ridiculous greening, how could I possibly romance him when my complexion resembled the inside of an eighteenth-century teapot?

With a sigh, then, I meandered over to the less offensive glow of my laptop screen to type up a blog post. After doing some preliminary research on compact fluorescent light (i.e., typing "compact fluorescent light" into Google), I found a website called One Billion Bulbs, which is trying to get people to log in with the number of bulbs they've changed to CFLs. Upon filling out my information, I got this cheery message in return:

"Thank you for taking the time to change 4 light bulb(s)," it said. "Your total annual savings are estimated to be $24.12. Your actions will help prevent approximately 491 pounds of greenhouse gases from being spewed into the atmosphere each year."

Aww—professional, and yet not afraid to use the word *spewed* . . . my kind of organization. And yet, $24? That's it? That's almost half a month's cable. If only I could just buy Mother Nature a few beers and call it even.

APRIL 10, DAY 41

Eat only free-range, organic, hormone-free (and if possible, local) meat, restrict to once per week

I've had an on-and-off relationship with Peter Singer since college. Not literally speaking, of course. But the author of *Animal Liberation* crept into my life sometime during the tail end of high school and has lingered in my heart and psyche ever since—sometimes prominently, other times in the background. Our longest stint together was about four years, when his ideas about animal suffering first struck me and I found it increasingly difficult to eat steak without tasting the blood, or eat pork without tasting, well, flesh (I maintain to this day that all pig tastes like human—no, I haven't eaten human, but I do know that these animals are as genetically similar to us as it gets; think of this next time you bite into a ham sandwich and I guarantee it'll be your last). My first dalliance with Singer, however, came to an abrupt end when I got a little drunk and cheated: I was at a bar, and this charming twenty-one-year-old engineering student put a bottle of cold beer in my hand and a basket of hot chicken wings on the table. There was no resisting. I plunged my hands into the saucy bits of deep-fried carcass, pulled out one of the meatier-looking morsels, and took a big, sinful bite. Surprisingly, I didn't feel much regret, and I spent the next few months gradually reverting to my carnivorous ways, first clandestinely at restaurants, then openly and recklessly at the grocery store.

But then, a year later, when I signed up for a philosophy class in moral issues, Singer and I bumped into each other again and I was forced to confront my feelings in a more academic way. After intellectualizing the various arguments for and against animal suffering, it wasn't long before I was running back into his warm embrace, begging for forgiveness and longing for that familiar world of veggie stir-fries and monthly PETA donations.

And so it went, back and forth, until now. The only constant in my pseudovegetarianism has been my aforementioned aversion to pork and things like headcheese or tongue—I worked at a deli once and would often have to slice jellied tongue for customers. It was the worst meat to slice because the bits would always fly right out of the slicing machine and land everywhere: on the walls, the floor, in people's hair. One time, a rather large hunk of cold tongue happened to fall directly into my cleavage. Needless to say, I couldn't help but feel a little violated.

These days, while I know for certain that I don't want any animal to suffer, I also believe in small-scale, family-operated farms with cows grazing in the fields and chickens running around spacious coops. When it comes to the abattoirs, I agree wholeheartedly with author and activist Temple Grandin that animals should be slaughtered quickly and humanely without being transported long distances and made to walk up ramps with the smell of death and fear everywhere. I think it's natural to eat eggs and dairy, too, provided it's hormone-free, preferably organic, local, and sustainably raised.

Still, I don't eat very much of it—probably only once a week, tops—as I understand the toll it takes on the environment, from the methane in cow farts to the nitrogen in chicken poop to the acres of land and gallons upon gallons of water required for not only the animals but their feed.

So to make a long story short (too late): I'm officially restricting myself to meat that's derived from animals who have been treated with respect, not antibiotics. This won't be an easy thing to explain to a waiter, and there's no doubt people will think twice before inviting me to their dinner parties for the next ten months—"Are you *sure* you want Vanessa there? She's become so . . . *difficult.*"—but at least I'll be able to reconnect with Singer a bit and maybe get into some heated arguments about ethical eating with all the vegetarians who frequent my blog.

APRIL 17, DAY 48

Invest in permanent laser hair removal rather than shaving and waxing

Journalists and publicists don't make good friends. The reason for this is that reporters are constantly being inundated with press releases, mostly about utterly boring and not-even-close-to-being-newsworthy events, sent by publicists, on behalf of clients who are desperate for coverage. But in the four years I've been working, I somehow managed to befriend one, a woman I can trust who never harasses me about lame stories. The funny thing with Sarah, though, is that she's so good at her job that she'll make a TV documentary about ancient cathedral architecture in Italy sound like the most entertaining thing to hit the airwaves in years—so before I know it, I'll go from interviewing Brad Pitt on the red carpet to researching the history of St. Peter's Basilica in the library.

The other day, Sarah left a frenzied message on my answering machine at work, run-on sentence after run-on sentence, and the only words I could really pick up were *photo shoot, Margaret Atwood,* and *blog.* I called her back.

It turned out she was at a photo shoot for a local magazine, which was profiling celebrity green couples in Toronto for its next cover story. Renowned author Margaret Atwood, who is a member of the Green Party of Canada and fairly vocal about her opinions on environmental issues, was there, so Sarah went over to talk to her while she was having her makeup done. At one point, my beloved publicist-turned-girlfriend mentioned Green as a Thistle and the 365-day challenge.

"Margaret was so impressed," said Sarah, "she asked for the URL, so I wrote it down for her and she said she was going to check it out as soon as she got home!"

"Really?" I said, in a mild state of shock.

But whatever sense of pride I might have had knowing that Margaret Atwood was going to be reading my blog quickly got lost in a wave of panic as I recalled today's entry: "Lasers before razors," all about laser hair removal.

As I explained in the post, I'd recently made the decision to get this treatment on my bikini line and underarms because it was permanent and meant no more shaving or waxing, which translated into fewer products like shaving cream, razors, wax, and canvas strips being consumed. It cost almost $1,000, but in the long run, it would be worth it. I couldn't afford to get my legs done so I'd still have to shave them every now and then, but my leg hair was pretty fine and grew slowly, so it was less of a concern. The laser machine would use some electricity, but it takes only five minutes of zapping and no more than eight sessions before the entire treatment is done.

The idea of never having to worry about unsightly stubble in my armpits and on my bikini line was all very exciting for me, of course—but did I really need to tell Margaret Atwood about this? It's so easy to write this blog and just disassociate myself from who, exactly, all my readers are, but now I had to stop and think about this. Not only was an award-winning author being inundated with my vain drivel on environmentally friendly hair-removal techniques, there was also my boss, my grandparents, ex-boyfriends, and hundreds of people I've never met. I might as well be walking up to strangers on the street, holding up my underarms, and saying, "Hey, do you have a few minutes?"

I logged back onto the site and went back into my post to rewrite the first paragraph, tinkering with it a bit more until it had some polysyllabic words and a semicolon or two. It was hardly Booker Prize–winning material, but at the very least I could claim that I may have gotten Margaret Atwood to contemplate the idea of firing a laser at a woman's bikini line, and in a bizarre kind of way, that was just as great an accomplishment.

APRIL 19, DAY 50

Turn off my freezer

A bottle of gin and a bag of peas.

That pretty much sums up the contents of my freezer right now. I've just been reading some stuff online about how the more food you store in the fridge, the more efficient it is—and something tells me that a solitary bottle of Bombay Sapphire and half-empty bag of organic sweet peas isn't helping me get a passing grade when it comes to efficiency.

There isn't anything else I want to put in my freezer, but then I couldn't possibly live without it, could I? Then, after more online meandering, I came across another green blogger by the name of Greenpa, with a site called Little Blog in the Big Woods. According to the introductory blurb, he's a fifty-nine-year-old grandfather living in a cabin in the woods somewhere in the United States and has been greening his life in numerous ways for the past thirty years; one of these, he says, is "No refrigerator . . . and you don't need one, either." Greenpa firmly believes his fridgeless eating habits have contributed to a healthier, more sustainable lifestyle.

This is great and all, but as I read further about his background and radical opinions, I began making that spirally motion with my index finger next to my ear because it sounded like typical crazy hippie talk. Even if he was right, that doesn't change the fact that he gets to live all Thoreau-like with his various eco-contraptions, and some extra time on his hands, while I'm living in an apartment with high-end appliances and a very low supply of patience.

But then I noticed that, on the inside of my fridge, there were two dials—one for the temperature of the fridge and a separate one for the freezer—again, not the fridge, only the freezer. So, presumably, if I turned the former down and the latter off, I could probably still keep all my perishables cool, just not cold.

And thus it was, I decided to turn off my freezer. I poured what was most likely a quadruple-gin martini, threw the remainder of my peas into a pot of boiling water, tossed the ice cube trays into the sink, and let the freezer door swing fully open to properly air out before it got warm. Done.

As I sat there at the kitchen counter sipping the last of my frigid alcohol, already feeling the rush through my temples and resentment in my liver, I glanced over at the barren expanse that formed the inside of my freezer. It looked like that abstract white-on-white painting by Kasimir Malevich—completely stark and uninspiring. And yet, something in me did feel inspired, ready to start cooking fresh food and actually eat leftovers instead of letting them succumb to freezer burn, ready to save money on the electricity bills, ready to stick it to Maytag, get up on the soapbox, and prove that the icebox is overrated.

APRIL 20, DAY 51

Pick up litter whenever I see it

There's this odd fellow in Toronto, about my age, named Mark, but most people just know him as the Litter Guy. He walks around the city all day, picking up trash from the street and putting it in a garbage bag. There's a big handwritten poster affixed to his back that explains things: he used to be a street kid but decided to do more with his time than just sit and beg for spare change. He roams around just about every neighborhood, stooping over to grab whatever lies in his path, except for maybe niggling things like gum wrappers or cigarette butts, and he accepts donations but doesn't ask for them. It's great that he's taken it upon himself to not only keep his city clean, but to make others realize that no matter how many sanitation crews the municipal government sends out with motorized suction carts and gas-powered garbage sweepers, we still manage to create enough litter for a guy like Mark to spend days picking up trash—in fact, there's so much of it, he'll often spend hours picking away without traveling more than a block.

But I'm not about to reprimand the City of Toronto—we actually have plenty of green initiatives in place, such as the weekly collection of recycling and compost bins (garbage pickup is restricted to every other week). Lawn clippings are gathered throughout the spring, summer, and fall, and are then turned into mulch and redistributed back to anyone who shows up at one of the many environment days held by councilors in their respective wards during the year. Then, there's the more recent project: the 20-Minute Toronto Makeover. All residents are encouraged to stop what they're doing at two p.m. on April 20, grab a garbage bag and a pair of gloves, and do what Mark does all day: pick up litter for twenty minutes. You can even get a free litter-picking kit from city hall.

Naturally, I figured this would make a great green change, but of course it couldn't be just a one-hour commitment—I'd have to pledge to pick up whatever litter I saw from now on, regardless of where I was or what I was doing. This, to me, sounded like it might be annoying but at the same time completely doable. Litter, after all, mostly just consists of crunched-up newspaper, empty coffee cups, and the occasional leftover sub sandwich (at least that's what people in the movies always seem to find in dumpsters). So I wrote my blog post, went in to work, and informed my editor that I was truly sorry but the story I was writing for Monday would have to be filed after deadline because the mayor said I could go pick up litter at two p.m. for half an hour.

"Do you want to come?" I asked.

He declined, but my colleague Maryam overheard our conversation and volunteered to tag along. Then Genevieve, one of our copy editors, suddenly rushed over and, with a flurry of hand gestures, explained that she loved picking up litter.

"My parents own a campsite," she said, "and when I was little, it was my job to walk around and pick up anything I saw on the ground—my dad was really strict about keeping it all perfectly spotless so we had to pick up every single sunflower seed shell and cigarette butt. Now it's just in my system and I pick up litter wherever I go!"

Works for me. So we gathered some leftover plastic bags from coworkers' lunches, stole some dishwashing gloves, and made our way out the back of the building to the parking lot. This area and, in fact, the whole neighborhood, if it can even be called that, is suburbia at its worst. It's a place where the road running between Leslie and Don Mills is called Lesmill, as though originality is just some mythical downtown phenomenon. The nearest restaurant is a place called Tako Sushi, located within a grayish brown cement block, which is bordered by the railroad tracks on one side and the parking enforcement division of the Toronto police on the other. It's not unusual to see a Canada goose or two milling about these parts, bobbing and pecking between Honda Civics and Ford Tauruses, looking entirely unimpressed (which, admittedly, is the way they usually look). The point is, while it was nice to breathe some fresh air, the environment wasn't exactly a refreshing change from the gray carpets, fluorescent lights, and vertical blinds inside, and although we were all looking forward to some exercise, picking garbage wasn't quite in the same category as a spinning class or hot yoga. Call it restrained enthusiasm.

We got to work right away, thanks to all the litterbugs at the *Post* who clearly lack the peripheral vision to see a garbage can right by the door, and by the time we got to the parking lot's entrance we had nearly filled our bags. Things were going smoothly and off in the distance, at the end of the street, we could see a group of other people with gloves and garbage bags doing the same thing as us, which made me smile as words like *community* and *civic pride* began floating around my head. But right at this moment, it happened—something so repulsive, so gut-wrenching, it would scar me for life: I bent over to pick up a discarded Pepsi can, lifted the thing up, and it was heavy. Like, full-of-a-solid-instead-of-a-liquid heavy. I gasped and dropped it, shaking my hand, suppressing my gag reflex and trying not to think of what could possibly be in there—a dead animal? Poo? A dead animal that died in its own poo? In a can? What else could it be? How did that heaviness even get in there? I couldn't suppress these thoughts, and I steadfastly resolved from that point on to pick up only litter that didn't have orifices. Also, nothing wet, rotting, or potentially alive.

APRIL 29, DAY 60

Spend part of each day learning about environmental issues

The Green Living Show felt like a green plague, a maurading wave of consumerism washing over me and drowning out almost all the motivation I had to keep this challenge going. It was meant to kick off today's pledge of daily eco-education, but after three straight hours spent plodding up and down every cramped row of an enormous conference center in Toronto's West End, stopping at all the booths and exhibits, nodding slowly and wide-eyed at the vendors and entrepreneurs and designers as they carefully and a little too slowly explained why their product, service, and/or business was eco-friendly and sustainable and socially conscious and vegan and LEED-certified and on and on and on, I felt mentally bloated, completely greened out, and sick of the whole thing.

Maybe it was the shamelessly overt marketing and sell-sell-selling or the throngs of modern hipster-hippies with their Bugaboo strollers and tote bags elbowing one another to grab as many free samples of organic granola and biodegradable vegetable wash as they could in one swipe. Or maybe it was the fact that my media pass was not only made from recycled paper with wildflower seeds embedded into it but also attached to a shoelace excitedly proclaiming, "I used to be a pop bottle!" Or it could have just been the fact that everything was green, quite literally, in color, and there was more than an occasional waft of body odor and hemp.

Whatever the case, it was all too much.

One possible reason for this might relate to my previous observation about the earnestness of New Age hippies—they're often so intent about meeting X, Y, and Z standards when it comes to greening their lifestyle but because they couple this with so little skepticism, it comes off as flaky, or even cultish. Then there are all the people who drone on about how amazingly eco-friendly a new product is for the environment, and how this-and-that technology can lessen your carbon footprint in ten different ways, but then they hop into their minivans and drive back to the suburbs (the parking lot at the Green Living Show, by the way, was full of cars—and they certainly weren't all hybrids).

This isn't to say I'm superior to these folks. In fact, if anything, I feel like just as much of a hypocrite, if not more, telling the world about how great I am for using recycled paper towels before taking a twenty-minute shower with extra hot water, changing into clothes that are far from meeting the ethical standards advised by Naomi Klein in *No Logo* and driving nine miles to work and back every day.

Speaking of recycled paper towels, I was about halfway through my rounds at the conference when I noticed the Cascades booth. They were taking up a lot of ground-floor square footage—about three times the size most booths were allotted—and had dozens of their products out on display. I thought it would be a good idea to stock up, so I walked over to the counter and asked if I could buy a two-pack of the unbleached towels.

"No, sorry," said the woman manning the booth. She was dressed in a severely angled navy blue power suit. "These are just for display purposes."

I didn't get it.

"But there are tons of them here," I said. "And this is the company that sells the product, and I'd just like to buy one pack. Please?"

"No, sorry," she repeated.

I still didn't get it.

"Hold on," I said. "You're telling me that Cascades put together all the money and resources and time and effort to come all the way here, and now you're not even selling your own product?"

The woman just shrugged and sort of grimaced.

This was ridiculous. What was the point of a Green Living Show if I couldn't even buy some stupid paper towels? And how could a green-minded company like Cascades just ignore the carbon cost of hauling all their merchandise, press material, and sales people from their head offices in Quebec down to Toronto to take up three booths' worth of space and electricity, all for the sake of putting in some face time to promote—but not sell—their product?

I could understand an organization that specializes in eco-friendly funeral services not having biodegradable coffins readily on-hand, or a company like Zipcar not being able to have actual cars available on-site (although, to their credit, they did). But this wasn't a standard corporate trade show, it was an event geared toward consumers.

After this, I decided to at least keep walking until I found a booth offering tangible goods for sale. It wasn't long before I saw, off in a corner, a woman in a folksy quilted apron selling what appeared to be nothing more than beeswax. On display was a pyramid of little jars full of the stuff, and a sign marketing the product as a thick, rich, natural alternative to moisturizer.

"Here, try some," said the woman. "It's great for eczema!"

Did I have eczema? I wasn't sure, but the backs of my arms were pretty dry, kind of bumpy feeling, and somewhat florid, so I took a wooden stir stick and plunged it into the tester jar. It was the consistency of raw honey (it probably was raw honey) and took a lot of friction before my skin gave in and absorbed it.

"It's great for eczema!" I heard the woman repeat to a new customer.

"Really?" came the reply. "You know, my sister-in-law has that, do you think this would work for her?"

"Oh yeah, it's great for eczema," said the beeswax lady, clearly on autopilot, and yet saying it with just as much enthusiasm as the first time, "and any other topical skin irritations, dermatitis, or any kind of rash. How long has she had it for? Does she use any natural or homemade skin-care products?"

"I'm not sure, but she's tried *everything*," said the customer, rolling her eyes and looking severely pained on her poor, overly dry sister-in-law's behalf.

Had I just stepped into an infomercial?

"You know," said the beeswax woman, "there are so many chemicals out there today, our bodies just can't handle the toxins anymore—and neither can the environment."

At this point, I couldn't handle the conversation, nor the fidgety toddler behind me pushing her way to the front of the beeswax counter, nor the screaming baby who'd probably just soiled his organic cloth diapers, so I left the Green Living Show with the most eco-friendly purchase I could have made: nothing at all.

APRIL 30, DAY 61

No more nonstick frying pans

You know what's easy to give up? Nonstick frying pans.

You know what isn't easy to give up? Driving.

I have the cutest car. I know, all people say that about their cars, but mine's definitely, objectively cute. She's a dark blue 2000 Volkswagen Beetle with a Turbo engine and light tan leather interior. I named her Bluebell, but I usually just call her my Bugaboo. She gets me to work and back every day, takes me grocery shopping, drives me to my parents' house on the weekend, up to Meghan's cottage in Haliburton, or to see all my friends in Montreal. She makes it through rainstorms, snowstorms, and that piercing Canadian cold. I've cried with her, sung songs with her, kissed in her, nearly gotten killed in her, and yelled, cursed, and slapped her across the dashboard. I've run her into the ground, rammed her into hard objects, and once nearly starved her to death, but I've also found her when she got towed and taken care of her when she was sick (which was my fault, in hindsight—I'd neglected to keep up with the oil changes).

The problem is, owning my Bugaboo is much like being in a sorority: I get solid companionship, but at a steep cost. Not only is there gas to pay for every few weeks, there's also the insurance, license plate renewal fees, emissions tests, maintenance costs, wiper fluid and other extras, parking, parking tickets, car washes, and on and on. But above all this, there's the environmental impact of having a car, which is huge. As I get hyperaware of every little detail in my life and how green or un-green it is, I'm starting to feel like a bigger and bigger fraud for owning a car. While I managed to survive without one for years, it's one of those things where—a bit like tasting real champagne—once you get it, it's hard to give up.

But then, as I was browsing through the program guide to Hot Docs, a local documentary film festival in Toronto that I'd just been assigned to cover for work, I noticed a short film called *To Costco and Ikea Without a Car.* It was by a first-time director named Peter Tombrowski, who had been living in the suburbs of Calgary without a car for nine years. Using a home video camera, he documented a recent expedition—on foot—to this big-box store miles away from his house, adding in some punchy music and more than enough irony, then self-published a book on the whole lifestyle called *Urban Camping.* I decided to order it online, then called him up to chat under the pretense of an interview, and asked whether he thought it was worth taking the plunge and becoming a full-time pedestrian.

"All right, Peter," I said, when I got a hold of him one afternoon at work. "Honestly, what made you get rid of the car? I'm guessing it wasn't the environment, at least not nine years ago."

"No, it started with the birth of our first child," he said. "Oh, and debt. My wife and I felt this need as new parents to do something different with our lives and take on more responsibility, but we'd also just moved and suddenly found ourselves cash-strapped, and the fastest way to get money was to sell our truck. We were definitely scared at first, but eventually, that fear turned into a kind of freedom."

"Do you ever regret giving it up, though?" I asked.

"No," he said. "I'm always hearing about accidents, traffic jams, road closures—those things are never an issue if you're walking. It takes a little more preparation and equipment—you actually need mountaineering gear, like Gore-Tex and hiking boots, so that's what we have. It's funny, we look as if we're off to conquer a mountain but we're just going to the video store."

That's not even close to being funny, I thought. Seriously, the last time I wore a Gore-Tex jacket was in sixth grade on a school camping trip. No movie is worth a fashion sacrifice like that.

But I moved on to another question—specifically, what Peter thought was the best part about living car-free.

He responded with that word *freedom* again, then mentioned how he'd become a better husband and parent. Freedom? Parenting? He really is delusional, I thought.

"Even better, though," Peter added, "is that whenever we're walking somewhere, the pace is always slow enough that we can stop and talk, and look at the world around us. There's more interaction, and we've become more reflective in general."

"Hmm," I said, still skeptical. "So what's the worst part? Not being able to get somewhere fast enough? Having to wear Gore-Tex?"

There was a pause, as though he wasn't sure whether to laugh or feel deeply insulted.

Maybe he really likes practical outerwear.

"Well, I think the heightened awareness can sometimes be aggravating and depressing," he said finally, "especially when we're the only ones on the sidewalk, when there's absolutely no one in front of us or behind us and a hundred cars passing by. We feel a little odd sometimes. You have to learn to be more patient and not so self-conscious. Cars are a very visible and significant part of our landscape here; they define us, they're a status symbol. So we have to get over that."

Peter then recounted his "microwave story" about the time he needed a new microwave and could retrieve it only by foot. Apparently, after going all the way to Sears, then making a grueling trip back home, dragging the appliance behind him on a dolly, he tried to install it but found it was too wide for the space, so he had to go all the way back and get a different model. Then, after lugging the second one home, he opened the box and saw it had a dent, so it was back yet again for a third time.

"The biggest trip for me, though, was when I went to IKEA for an office table," he said. "It weighed one hundred and twenty pounds, and I actually carried it. I had straps that were wrapped around the table, and other parts in my backpack. We also used to buy chairs one at a time from IKEA—it took a year and a half to get all eight in the set."

A year and a half? This was not cool.

I asked about cocktail parties and other fancy events, especially with regards to his wife—I mean, how could she walk all those miles in heels and a pencil skirt? Surely no pair of running shoes could ever fit in a clutch purse.

Then came the dreaded reply.

"She brings a backpack," he said. "Then she'll change into nice clothes and shoes when she gets there."

This no-car thing was looking like less and less of a possibility. But then Peter said something else, right before we hung up, that made me reconsider everything, especially as I was sitting in traffic at the end of the day on the highway, alternating between gas and brake, gas and brake: "We have to remember that our first big accomplishment in life is learning to walk. Physically, as humans, we are made to walk—not push a gas pedal."

For whatever reason, this evolutionary rhetoric struck a chord with me. Peter was right—I had a nose with which to breathe, eyes in order to see, and feet that were meant to walk. Not having a car might inconvenience me in numerous ways, but just because this guy in Calgary happens to wear unflattering Gore-Tex and backpacks and make multiple trips to Sears for a single microwave doesn't mean I have to. It's definitely possible to wear a dress and ride a bicycle, more than possible to carry a few bags of groceries on the streetcar, and, even if I'm purchasing something at IKEA, there are car-sharing options or taxis.

Ultimately, I think if I'm going to take this leap, I'll have to not only keep thinking about all the money I'll save, but also convince myself that going carless is essential in fulfilling my biological destiny. This one's for you, Darwin.

may

1	No more Q-tips
2	Switch to natural deodorant
3	Recycle my wine corks
4	Give up chewing gum
5	No more makeup remover pads
6	Don't use my oven
7	Use natural bronzer
8	Buy only loose-leaf tea and reusable teabags
9	Switch to natural, organic shampoo
10	Fill kettle with exact amount of water needed
11	Shut down computer at end of day
12	No more air conditioning in the car
13	If it's yellow, let it mellow
14	Use matches rather than a butane lighter
15	Carbon-offset all my flights
16	Use herbs and spices grown at home
17	Unplug my refrigerator
18	Lobby, petition, and letter-write on behalf of environmental causes
19	Request the vegetarian option for in-flight meals
20	Shop for books at local, independent bookstores
21	Buy music only as mp3 files
22	Get allergy shots instead of pills
23	Give "green" gifts
24	Replace acne-ointment with tea-tree oil
25	No more food delivery
26	No more take-out
27	Request no receipt at stores, recycle those received
28	Refill all possible containers
29	Switch to an eco-friendly dry cleaners
30	Use a PVC-free shower curtain liner
31	Turn off the air conditioner

MAY 1, DAY 62

No more Q-tips

May 1st. May Day. Also my birthday. I just turned twenty-eight, decided to give up Q-tips in an easy, random, and purposely-unrelated-to-my-age blog post, then celebrated in the most unsocial way possible—a dinner at my parents' house with Mom, Dad, Emma, and Meghan. Truth be told, I wouldn't have wanted anything else. Actually, on second thought, it might have been fun to have a dinner with Mom, Dad, Emma, Meghan, and a Tall Handsome Man with Good Taste in Wine and Better Taste in Film, whom My Parents Liked but Didn't Like Too Much. Still, it was nice.

Despite my mother, of all people, forgetting that I was turning twenty-eight (she wrote "Happy 27th!" in the card and put twenty-seven candles on the cake), she at least didn't forget about my green challenge and made sure to cook food that was organic and free-range wherever possible. Meghan, who never got less than an A+ in high school and college, presented me with an A+ gift, of course, combining both of our current obsessions: health and the environment. It was a stylish tote bag with Mason jars full of homemade tea, toasted nuts and seeds, preserves, and hummus, each wrapped in leftover fuchsia crinoline from an old bridesmaid dress she had sitting in her closet.

The meal and gifts were all so thoughtful and truly warmed the green cockles of my heart (whatever those are). And yet because a part of my heart belongs to the Grinch, I couldn't help but feel like a bit of a caricature, too. When I was a kid, I went through a pig phase. I just decided out of the blue that I really liked pigs and went around telling people why they were my favorite animal as well as writing school projects on subjects like truffle-hunting pigs or the cleanliness of pigs, sketching Piglet and Wilbur and Porky over and over, and collecting various porcine paraphernalia from pig stickers to pig figurines to pig key chains. Eventually, I got sick of them, but it took years to undo this reputation—every birthday, every Christmas, any gift-giving occasion whatsoever, the only present people ever thought to give me was pig stuff.

So now I'm a little concerned that for the rest of my life I'll be accepting armfuls of green goods, as though anything that could potentially be in the least bit ungreen just isn't an option with me. I suppose it's better than pigs. But I'm really thinking that, when this challenge is over, I should make an effort to buy the occasional Gap shirt or can of hairspray, if only to remind people that there's more to me than tote bags and vegan dental floss.

MAY 2, DAY 63

Switch to natural deodorant

A comment I just received on my blog post about switching deodorant, from a person with the screen name Bureinato: "I've just drifted over to your site . . . and wanted to share that I've been using Listerine on my armpits and it's the best non-aluminum deodorant I've ever used."

I'm starting to get a lot of weird people visiting my site. Another woman left a comment recently in which she confessed an irrational fear of hair dryers, and more than a few readers have serious emoticon addictions, throwing a happy face, surprised face, sad face, and disgusted face all in the same comment. Wordpress, the program I use to manage my blog, also has a box on their stats page that tells me what term a person has searched for in Google that's led them to my site, and it's very often surprising, and sometimes flat-out disturbing. There are fairly normal searches like, "What is a thistle?," "natural deodorant," or "Vanessa Farquharson" (okay, maybe that was me). But then there have also been searches for "monkey clip art," "sexy dentist," "earth on a cake," and, my favorite, "Do chocolate cell phones have cameras?"

If only I could find a way to link my deodorant post to searches for Paris Hilton . . .

MAY 4, DAY 65

Give up chewing gum

"I don't get why people chew gum, anyway," said my sister. "It's, like, so exhausting. Who wants to move their jaw around chewing for an hour when you don't even get to swallow it afterward?"

She had a good point. Although, this also comes from a girl who thinks iron lungs should be made available for commercial use, and once wore my mother's old neck brace just for fun, claiming she was too tired to hold her head up.

MAY 6, DAY 67

Don't use my oven

After dropping about 240 percent of my paycheck on a buttload of stretchy pants at Lululemon this past September, I suffered severe posttraumatic-impulse-buy syndrome and, in the midst of my delirium, thought it would be a good idea to overcome this by draining another $40 from my bank account to join a group of Gatorade-chugging, PowerBar-binging adrenaline worshippers who planned on getting up at the crack of dawn on the first Sunday in May, squeezing into a variety of tight-fitting attire, much like the stretchy pants I had just purchased, hauling ass up to the intersection of Yuppie and Preppy, and then running ten kilometers alongside eight thousand other crazy morning people.

It's called the Sporting Life 10K, my newspaper sponsors it, and I was registered for it. For some reason, holding all that overpriced merchandise in my arms made me feel as though if I committed to this event and trained really hard for it, I could maybe become a runner, not just a jogger.

Now, I should say up front that I'm very much a firstborn child: my parents' greatest fear—it seemed to me, at least—was that I might grow up and not be proficient at something; their way of preventing this was to enroll me in every day camp, after-school program, and sport ever offered in this city from 1982 to 1995. At one point or another, I've been a swimmer, a figure skater, a squash player, an equestrian, a tap dancer, an indoor rock climber, a tennis player, a ballerina, a long jumper, a martial artist, and a skier of the downhill, cross-country, and water varieties—and yet somehow I've never been a runner. It's never mattered much, but recently, a lot of my friends started registering for half-marathons and, because I can barely make it around the block these days without collapsing in a florid, wheezing mess, I became determined to keep up with my friends and at least run ten kilometers before I die.

So I began training, following the schedule recommended on the Sporting Life website down to a tee, checking off the routines each day on my calendar. Surprisingly, it went pretty well. I suffered only the occasional bad day of cramping or exhaustion and managed to avoid full-out cardiac arrest. Bonus! Unfortunately, though, while my physical state was keeping up, my mental state was not; it wasn't long before I found myself fretting—not fretting about, say, the potentially uneven weight distribution on my arches (I had already put into practice various ways of counteracting any pronation), but rather the ecological footprint I was leaving behind. I figured running would be pretty low-maintenance and easy on the earth—no gallons of chlorinated water required, no rackets, clubs, or safety gear; plus, I'd sworn as part of this green year not to use any treadmills, so the only energy being expended was my own. However, what I didn't foresee was the amount of laundry that would pile up after every outing: a pair of sweaty socks, underwear, a T-shirt and long-sleeved shirt, plus a sports bra would all need washing every time I went out. On top of this, becoming a serious runner meant printing out training schedules and investing in new shoes, a new windbreaker, special earphones that wouldn't fall out as I ran and a belt holder for my iPod, a squishy water bottle, a tight-fitting headband to keep stray hairs at bay, and retro terry cloth wristbands to look extra cool (all right, I didn't actually get wristbands, but I was close).

As the weeks went by, I gradually figured out a system to minimize the laundry pile, using the previous workday's socks and shirt, and reusing shorts two to three times before washing them. I also began making my own energy-boosting drinks and resisted the urge to buy a heart-rate monitor. But just as I was convincing myself that this form of exercise might be justified as eco-friendly after all, the time came to pick up my race kit. It was the day before the actual run, and the kit consisted of an envelope with my time chip, an ugly T-shirt, free samples of sunscreen, energy bars, and other overpackaged stuff I really didn't need, and it all came in an enormous plastic bag. A bad sign. But maybe the race itself would prove to be more minimal.

The next morning, after rolling out of bed twenty minutes late, scrambling into my clothes, and downing some oatmeal and coffee, I hailed a cab up to north Toronto, picking up my work friends Maryam and Justin, who'd also signed up for the race, on the way. The only reason I agreed to share a cab with them and suppress my new-at-running insecurity was because Maryam had just pulled a calf muscle in a ballet class and Justin's breakfast consisted of three cigarettes.

"I didn't even know you were doing this race until Maryam told me this morning," I said to Justin, whose neon orange Adidas jacket seemed at odds with his hangover. "How are you feeling?"

"Oh, I'm all right," he said, with a sigh. Justin has this funny way of speaking—almost everything he says is conveyed with a wistful, singsong tone of voice.

"So when did you sign up for this?" I asked.

"Um, yesterday?" he said. "Yeah, yesterday."

He explained that it was his summer goal to get in shape, but that the only way he could force himself to do so was by signing up for stuff like this.

"Gotcha," I said. "And you, Maryam? How's your ankle?"

"Well, it hurts, but whatever," she replied, with typical nonchalance. She'd done the race a few times before, not to mention a half-marathon the previous fall, and wasn't concerned with beating any personal records.

"How are you feeling?" she asked me.

"I think I'm somewhere between fainting and barfing," I said.

And so it was that Maryam and Justin spent the rest of the cab ride reassuring me that I'd do just fine and the whole thing would be over soon anyway.

After arriving and checking in at the sign-in table, we all went to the bathroom, then warmed up and stretched, took our places, and waited for the gunshot.

Noise pollution, I thought to myself.

But as I crossed the start line and began coasting down the longest street in the world, surrounded by the rainlike sound of eight thousand feet hitting the pavement, my eco-anxiety was relieved and nervousness turned to excitement.

About two miles in, however, excitement turned back into nervousness, then fatigue, and, by the halfway point, I just wanted to finish the damn thing and get some brunch.

As well, it was difficult to appreciate the freedom of a car-free run and clean air while Mother Nature was getting trampled in the wake of promotional swag, discarded wax-paper cups, and porta-potties. Then, at the finish line, medals were handed out to everyone, a family-friendly rock band plugged in their generators and amplifiers, and organizers in various tents began handing out bagels, bottles of purple-flavored high-fructose corn syrup, and bananas imported from Colombia. If it weren't for the fact that I finished in less than an hour and was on a blood-rushing high, I probably would've felt a lot more irritated.

Maryam and Justin, both of whom by this point had cooled down and were hovering around the results board—Justin's cigarette breakfast had actually seemed like a smart choice as he came in at 52 minutes, while Maryam finished in 57—saw me wandering toward them and came to offer their congrats. Then I found my half-asleep-but-still-beaming mom, dad, and sister, who gave me a round of hugs. My dad offered to retrieve some bagels and Gatorade, but I shook my head and politely declined as both my stomach and my conscience were screaming out for some free-range eggs Benedict with organic English muffins. Seeing as I'd switched off my oven for the rest of the year and there'd be no more baked goods unless someone else was baking them, I decided a local restaurant with a sustainable menu was the answer to my protein and carbohydrate needs—the only problem would be convincing my legs to start moving again.

MAY 9, DAY 70

Switch to natural, organic shampoo

Lurking somewhere in a six-ounce plastic bottle, sitting somewhere on the crowded shelf of the personal care aisle in a drugstore, there's a magic potion that will turn my limp, frizzy, mousey-brown hair into lustrous chestnut waves. It's been there all along, I just haven't found it yet—but one of these days, I'm going to try yet another shampoo brand, and right when I'm least expecting it, I'll step out of the shower, blow-dry my hair, look in the mirror, and there it will be: cue the swelling chorus of *Hallelujah* as I toss my head from side to side, run my fingers through this newfound silky mane, and throw all my hats, bobby pins, and ponytail holders in the garbage because there will never again be a bad hair day.

This has been my logic since I was about thirteen, when my weekly allowance became enough that I could start buying my own shampoo. It's ridiculous, yes, but I'm sure plenty of other women harbor the exact same delusion, or there wouldn't be over a hundred different brands of what is essentially just a slight variation on soap available on the market.

The more I learn about toxins and chemicals, though, the more skeptical I am about brand names. Not all of them are carcinogenic and evil, but most of them at least capitalize on our desire to look, smell, and feel pretty. This translates into artificial dyes (yellow = lemon), fragrances (#54 = apple), and additives like sodium laureyl sulfate to make things foamy, or antihumectants to make things smooth. In turn, this means that when I snap open a bottle to take a whiff of my "morning spring" scented suds, then rub it into my scalp until lots of bubbles appear, maybe even rinse and repeat, it cleans my hair in the short term but could also do some damage to either myself or Lake Ontario in the long term.

In light of this, I decided to switch to a natural shampoo. But I didn't want to buy a whole bottle only to discover it was useless, so I took some leftover hotel mini-bottles I had from my last vacation and went to the Big Carrot, where there are bulk dispensers of at least five or six different eco-friendly brands, all of which are paraben-free, SLS-free, preservative-free, and in some cases even fragrance-free—although certainly none of them were marketing-free, boasting similar claims on their labels of pH-balancing, replenishing, and hydrating formulas for oily, highlighted, dry, or normal hair. But of course, being someone who's more than happy to be marketed to, I pumped out samples of each one and took them home to experiment.

The empirical results were, in no particular order: lousy, yucky, stinky, runny, and blech. The third one actually made me stop and double-check the ingredients online afterward because it seemed as though maybe I'd bought moisturizer by mistake, and I'm pretty sure the fifth one actually made my hair dirtier.

It was thus with a mess of greasy tangles framing my rather annoyed and puckered face that I wrote a blog entry about the natural shampoo switch, and I certainly didn't hold back on the complaints. As it turns out, my readers had plenty to say, too. Some recommended brands they liked—Giovanni, Nature's Gate Organics, and the Lush shampoo bars—but more interesting was the dialogue that started up about this so-called No 'Poo movement, which is short for No Shampoo. All the followers of this were adamant that if I just stopped washing my hair, my body would figure it out, alter the amount of oil it produces, and eventually I'd have shiny, but not greasy, tresses.

The only catch was that it might take a week or so for my system to adjust. Unfortunately, this was just too big of a catch. I have to go into an office every day, and despite the fact that some of my male colleagues have been known to wear moth-eaten sweaters, dilapidated fuzzy slippers, and fake glasses around the workplace, I had somewhat of a stylish arts journalist reputation to keep up. There wasn't going to be any grease pit forming on my noggin anytime soon, no matter how temporary.

Going back to the blog, however, I noticed a few other women had left comments about spraying vinegar on their hair in place of shampoo. While not using anything at all was too extreme, using an everyday item from my pantry seemed completely doable—people have, after all, been using vinegar for years around the house as a natural cleaning product, so surely it would accomplish something hygienic—and besides, what would this whole green year be if I didn't at least eschew my vanity every now and then?

And so it was that I found myself standing in the shower, looking down at a ridiculous little row of products: there was an exfoliating lemon-verbena body wash, a nourishing chamomile conditioner, and . . . a jug of white vinegar. Because I had no spray bottle, the vinegar application consisted of simply unscrewing the cap and pouring it over my head. In the hot shower, it actually felt cold, which startled me—I gasped and jerked my hand, accidentally tipping a couple more glugs of the stuff out, which promptly careened into my eyes so it felt as though a steady stream of razor blades was flowing across my corneas. I dropped the jug, panicked briefly, and in choosing between saving my vision and saving the vinegar, decided the former took precedence. I let the last ounce run down the drain, rinsed my eyes and hair for another few minutes, and finally stepped out, feeling a sudden empathy for all cucumbers destined to become pickles.

On the plus side, my hair actually felt pretty smooth, but on the downside, I reeked like a fish-and-chips shack. My next bottle of vinegar would be staying in the kitchen cupboard.

MAY 13, DAY 74

If it's yellow, let it mellow

Here is what I will say about the "If it's yellow, let it mellow; if it's brown, flush it down" rule: this is fine, providing you drink a lot of water. And by a lot, I mean permanently attaching your mouth to the faucet. I mean that your kidneys feel overqualified for the job, you don't know the meaning of waterlogged, and Hydration is your middle name.

Conversely, one should not attempt this if one is anything like, say, my sister, whose daily liquid intake consists of coffee in the morning, the occasional swallowing of saliva in the afternoon, sometimes more coffee later on, and wine in the evening. If Emma were to pee for a day or two and not flush the toilet, she wouldn't so much be "letting it mellow" as letting it steep, and possibly even ferment. A lot of eco-minded people like to go on about how urine is clean, a perfect disinfectant. And, true, with its naturally occurring levels of ammonia, it's great for when you're on your own in the woods with no first-aid kit and get bitten by a snake (in which case, you're an idiot and deserve to pee on yourself). But if you're not well hydrated, let me tell you, two-day-old pee is not going to clean the toilet bowl; it's going to give it jaundice and stink up the whole bathroom no matter how many matches you light in there. The irony of all this is, of course, that in order to pull off the "if it's yellow" rule successfully, you need to drink almost as much water as it takes to flush the stupid thing each time, which is why I'm reneging a bit on this rule, flushing at least every other pee and before guests come over.

MAY 15, DAY 76

Carbon-offset all my flights

When it dawned on me that not quitting my job meant I'd have to continue working (how did that happen?), I decided this would be fine as long as I could take six weeks off. Totally reasonable.

So I decided to approach my editor one afternoon, after he'd had a few sips of his four p.m. Americano from Starbucks, and ask him—well, really, more like tell him—that I was going to take a minisabbatical. Or, technically speaking, an extended vacation. I'd already concluded that if he said no, I'd quit and do it anyway because I was just that desperate to get out of my fluorescent-lit cubicle in the suburbs. The problem is, I love what I do, and I love my colleagues, but a one-week package holiday in the Caribbean and ten days over Christmas at a hotel called My Parents' House is not enough to get me, nor my sanity, through the year. Fortunately, my editor is beyond accommodating and loathes the faintest glimmer of conflict or confrontation, so he barely even blinked before agreeing to my proposed leave of absence. (Of course, it might have helped that I'd planned to go during a slow time for the movie industry, arranged some backup freelance reporters to cover for me, and assured him I'd be back with renewed professional enthusiasm and an assortment of tacky souvenirs.)

Of course, my editor is not stupid, so he also managed to find a way for me to continue filing stories during this time.

"How would you feel about turning your blog into a column?" he wrote in an e-mail, a few minutes after I'd told him. "We'd call it My Green Year, it would run every Thursday, and you'd retain all the rights, of course."

Huh, how about that? The editor in chief wasn't interested in a green column, the managing editor wasn't interested, but my very own arts and life editor—to whom I clearly should have gone in the first place—had come to my side.

I wrote back, accepted the offer, and asked whether he was perhaps an environmentalist at heart, too.

"Well, I thought it would be an easy way to get more advertising dollars," he said.

This wasn't exactly the response I'd anticipated, although I did work at a conservative newspaper that was struggling to get out of debt. But then, I recalled that at my editor's housewarming party a little while ago, there weren't any paper towels, and his fiancée, with whom he ran a fashion blog, mentioned she was trying to restrict her clothing purchases to local thrift stores. Maybe, then, Ben was just a closet environmentalist.

Either way, this got me excited—a column and a six-week vacation. Suddenly, my job rocked! Still, my elation at being granted six weeks off work quickly morphed into stress. There was so much I wanted to do, so many places I wanted to go, and so many people I wanted to see—it was like being told, as a kid, that I could pick only one TV show to watch after school, one ice cream flavor out of the thirty-one available at Baskin Robbins, all of which I craved (except maybe Tigertail). And not only would I have to plan a vacation that had maximum experiential value for a minimum cost, I'd also have to figure out how it would fit into my green challenge. Sadly, if I ended up going to all the places I wanted to, there would be even less chance of justifying it environmentally than there would be of the *Post* moving downtown and investing in solar power.

But could I really let guilt stand in my way? How often would an opportunity like this come up? I sighed and looked down at the itinerary I'd drafted up after some flight searching and schedule tweaking. The summer vacation of all summer vacations, if I went through with it, would look something like this:

Toronto to London, England: Here, I could visit my friend Kelly, the former theater critic at the *Post* who was now working for the *Guardian,* as well as my oldest friend, Kate, who grew up on the same street as I did and was now in culinary school. My family was also planning a trip to England in July to celebrate my aunt's birthday in the Cotswolds and then head north to visit my grandparents up in Sunderland.

London to Ramallah, Palestine: After returning to London from Sunderland, I could then hop on a plane to Tel Aviv, cross over the wall in the West Bank, and see Jacob in Ramallah. He's been trying to convince me to visit for almost two years, and it's a part of the world I learn about every day but have yet to ever see. As well, if I do go to the Middle East, I want it to be at a time when one of my friends is there—someone who knows the region, the history, the languages, and both sides of the political divide. Jacob fits all these criteria. Not only is he the smartest person I know, he's also fluent in both Arabic and Hebrew and can throw together a mean road trip in less than forty-eight hours. It's a long distance to go, but I'm sure it would be worth it.

Ramallah to Ávila, Spain: While trawling a travel writers' website a little while ago, I came across this company called VaughanTown, based in Madrid. Despite its horrible name, it offers a remarkably cool service: Spaniards looking to improve their English pay to stay at a villa in the west country, where they spend one week speaking with people from Canada, the United States, England, and Australia. The Anglos are volunteers but have all accommodations, food, drink, and transportation to and from Madrid paid for by the company. It sounds almost too good to be true, but really, how bad could a free week in the Spanish countryside be?

Ávila to Portland, Oregon: On Treehugger.com, there's always this little ad at the bottom of the page for a "Sustainability and Energy in Motion Bike Tour." Again, not the greatest of names for an organization, but either way, it sounds interesting: a bike ride through the valleys of Oregon with a focus on sustainability. The food is all vegan and the riders learn about everything from organic dairy farming to composting toilets. I e-mailed the folks there and asked about a media discount. If that comes through, it'll be a perfect way to get some firsthand experience with the green community on the West Coast while also getting some exercise and meeting new people at the same time.

Portland to Muskoka, Ontario: If I were to spend ten days in England, five days in Palestine, one week in Spain, ten days in Oregon, and a few days in Toronto between flights, this leaves me with one week before going back to work. My parents were talking about renting a cottage up north in Muskoka at the end of August, which would be the perfect place to relax and do nothing. I could catch up on any blog comments I may have missed, think about new green changes, and get myself together before the insanity of the Toronto film festival hits in September.

If this all comes together without a hitch, I'm thinking it's worth considering a backup career as a travel agent. Looking over my potential itinerary, it's at once ludicrous and brilliant, indulgent and exhausting, and it's probably the most flying I'll ever do in such a short period—17,270 miles and 349 gallons of fuel, which amount to 6,828 pounds of carbon dioxide and at least a million tons of guilt. My hypocrisy is once again slapping me right across the face: how on earth can I purport to be such an eco-minded hippie intent on leaving a lighter footprint if I go ahead and spew all this pollution into the air for the sake of meeting up with a couple friends, seeing a few new places, and reacquainting my skin with its long-lost frenemy, the sun?

My 365 cumulative green changes won't even begin to undo the carbon repercussion of 365 seconds in a jet plane.

But the more I think about it, the more I realize that despite all the ways I could potentially get around this air travel dilemma—using my Internet connection and a webcam to see family and friends, teaching English here in Toronto, reading about the Israeli-Palestinian conflict in the newspaper—it just isn't the same. I can't celebrate my aunt's fiftieth birthday online, I can't virtually hug my best friend or imagine the scale of the wall running along the West Bank, and I can't just read about how great a freshly picked organic plum from Corvallis, Oregon, tastes. Some experiences in life are meant to be visceral, and the further I travel from my safe, regimented life here in Toronto, the more direct, immediate, and powerful my experiences will be.

I'm reminded of Al Gore and all the criticism he endured for flying to a new city or country practically every two days in order to give his *Inconvenient Truth* presentation—while it's true he probably didn't need to go to every single one of those places, on the other hand, what if he just sat at his desk the whole time giving phone interviews and hoping that his slide show got featured on YouTube one day? His voice wouldn't have reached nearly as many people, and the ripple effect of the green movement surely wouldn't have spread as far as it did. Obviously, my vacation won't be inspiring a global campaign to save the planet or anything, but the point is, a problem like this isn't easy to solve; the power of the green movement, the intellectual and social values of a transatlantic flight, and which people are effecting what levels of change are so subjective. For now, perhaps the best answer when it comes to air travel lies somewhere in the middle, somewhere in carbon-offsetting, which at least lets people make sure that a few trees get planted to balance out the pollution.

Critics like George Monbiot, author of *Heat,* the best-selling book about global warming, like to compare organizations such as TerraPass or the CarbonNeutral Company—which sell offset credits to both individuals and businesses—to the medieval church selling indulgences to absolve sinners. It's a bit of an exaggeration, really. I still think polluting and doing something is better than polluting and doing nothing, so with this lesser-of-the-two-evils mentality, I researched the specific projects at TerraPass, swallowed the remains of my guilt, booked all my flights, and offset them for $36.95, plus a complimentary luggage tag.

Mr. Monbiot, you can officially call me a sinner.

MAY 17, DAY 78

Unplug my refrigerator

I just unplugged my fridge. The whole thing. I know.

Here's what happened: when I turned off the freezer section a while back, it gradually began affecting the temperature of the entire unit, making it near impossible to keep track of just how quickly my perishables were perishing. Because I'm somewhat neurotic about the freshness of my food (there was a traumatic experience in my youth with expired chocolate milk—I'll say no more), and also because of a lack of ventilation and subsequent issues with humidity and odor, I finally realized I'd have to either endure another ten months of increasing frustration and guesswork, or just give up on the whole damn thing, so I pulled the plug. Nothing catastrophic has happened yet, but something tells me the regret I once suffered upon consuming an entire serving of General Tso chicken after a night of heavy drinking will pale—truly pale—in comparison to what I feel in about a week from now.

MAY 30, DAY 91

Use a PVC-free shower curtain liner

After checking Treehugger at least a few times each day to keep up with all the eco-news, I noticed there was a surprising number of Toronto-related posts; looking at the bylines, I saw that a writer named Lloyd Alter was behind most of them. Treehugger was being frustratingly unresponsive about adding me to their blogroll—that is, the list of other green sites they recommend in their sidebar—so I thought I'd try a little media back-scratching: I'd find a way to write about this Lloyd guy if he found a way to profile Green as a Thistle on Treehugger. A few days later, I had found my way in: a feature article for the weekend edition of the *National Post* about how green Ontario's capital really was and whether mayor David Miller's ambitious plans for LED streetlights and green roofs was actually feasible. Who better to ask than Lloyd? He e-mailed back, said he'd be more than happy to chat, and that yes, my website looked great, so he'd definitely write about it.

And so it was that I ended up sitting across from this feisty, five-foot-two architect-turned-blogger dressed in all black, sipping a fair-trade chocolate espresso at a table for three—one chair for him, another for me, and a third for our bike helmets—listening to his opinions on everything from the irony of environmentally friendly potato chip manufacturing:

"We have potato chip factories that use solar and wind power, but hello, shouldn't we just not eat potato chips in the first place? It's like greening an ammunition plant!"

to the importance of green leadership:

"David Suzuki is depressing and *An Inconvenient Truth* is really boring."

to shaming as a relevant tactic for environmental reform:

"People are embarrassed now to drive Hummers, which is good. I'm all for humiliation."

I was starting to zone out when the conversation turned to the economical advantages of prefabricated housing, but suddenly I thought I heard him say something about sperm.

Turns out, I heard right.

"You tell people their shower curtain might be lowering their sperm count, and suddenly you've got everyone listening," he said. I guess it was obvious I hadn't been listening.

He explained how the pthalates used in the manufacturing of polyvinyl chloride, the material with which most plastic shower curtains are made, has been shown in studies to damage sperm DNA. Meanwhile, PVC itself has been classified as a carcinogen and linked to increased risks of brain tumors, cancer in the spinal cord, and erectile dysfunction.

I wasn't particularly concerned about my sperm count, but the idea of inhaling the steam off a carcinogenic curtain every morning was unpleasant enough for me to pedal home as fast as I could, rip down my grungy shower liner, book a Zipcar, and head to IKEA for a PVC-free version, which managed to fulfill both of my purchasing requirements: it was cute, thanks to its Swedish moniker, *Näckten,* and it was cheap, thanks to its sale price of $1.99.

june

1	Recycle anything and everything that can be recycled
2	Switch to all natural, minimally packaged eye shadow
3	Switch to natural bar soap and a recyclable traveling case
4	Sleep naked
5	Use natural, mineral-based sunscreen
6	No more baths
7	Let my hair air dry
8	One bar of soap only for face and body
9	Order smaller, more eco-friendly business cards
10	No more nail polish
11	Water plants with graywater
12	Go to eco-friendly spas
13	Weed Mom's garden by hand, rather than with pesticides
14	Use gauze rather than Band-Aids
15	Switch to natural cat food
16	Take the stairs instead of the elevator
17	Use reusable cloths for household cleaning
18	Use natural hair dye
19	Raise environmental awareness with stickers, blogs, and other media
20	Use GoodSearch.com and Ripple.org for Internet searches
21	Cure hangovers without pills or plastic barf bags
22	Treat sunburns with pure aloe
23	Rent only hybrid cars
24	Cancel my subscription to a second newspaper
25	Sell my car
26	Wash dishes by hand
27	Use subway tokens rather than tickets
28	Buy cereal in bulk
29	Use vinyl-free, recycled-paper photo albums
30	Iron clothes only for special occasions

JUNE 4, DAY 96

Sleep naked

THE FIRST TIME I ever slept naked was in the fall of twelfth grade. I was dating Eric, a short, eccentric Jewish boy who rarely acted his own age. I don't mean that he was immature, per se—rather, he just seemed caught between wanting to be a child again and wanting to be a forty-five-year-old literature professor. A funny connection between Eric and my green year, by the way: his grandfather founded (and his mother at the time owned and ran) a company called NOMA, which supplies most of Toronto—and, in fact, North America—with electrical wiring and light bulbs. Although it was bought out by another firm some years ago, the NOMA brand continues to appear on everything from LED Christmas lights to the "soft white" compact fluorescent bulbs I now have by my bed, so there's a little bit of ex-boyfriend nostalgia every time I turn them on and off.

Anyway, back to sleeping naked: by the time I was seeing Eric, I was old enough to drive, but not drink, so my parents would let me take the car to go see him at night. We'd usually talk and fool around in his bedroom and eventually drift into sleep until about four a.m., when my internal panic-alarm would go off, at which point I'd get up, scramble into the pile of clothes on the floor, sneak out of Eric's parents' house, drive back home, sneak into my parents' house, and try to will myself into being a few pounds lighter as I navigated the creaky stairs up to my room on the top floor.

One of the things I always noticed while somnambulantly reclothing myself at Eric's was that he never got out of bed, even if he was naked. One night, I asked him, "Don't you want to get up and put pajamas on? Or boxers, at least?"

"No way," he groaned. "It's so much more comfortable without anything on. Nothing gets bunched up or twisted, no tags scratch at your neck. It's the only way to go. Seriously. Try it."

"But what about in the winter, when it gets cold?" I asked.

"Turn the heat up," he said through a yawn, and rolled over to go back to sleep.

One night, or rather morning, when I got back into my own bed, I decided to give it a whirl. Something about wearing nothing—not even a pair of underwear—felt kind of licentious, kind of bad. What if my father came in to wake me up in a few hours for breakfast and there I was, sprawled supine, naked and exposed on top of the duvet?

Still, it was worth trying. So after taking off all my clothes, I slipped into bed and pulled the covers up close to my chin, curling myself into a fetal position. It was weird, at first, feeling all the sheets against every part of me, but after a while it began to feel nice. And eventually, when I woke up again, it felt really nice—the word that came to mind at the time, in fact, was *natural.*

Throughout college, however, this habit became less frequent, and now I sleep naked only if it's a hot summer night (and that's hot in both senses of the word, I suppose, although it's referred strictly to temperature for the past year or so).

But while I was evaluating my laundry the other day, thinking about what I could do to reduce the volume of stuff being washed every couple of weeks, I realized my flannel pajamas were taking up a lot of space. If I were to go back to a routine of sleeping naked, I'd be able to save on detergent, water, and electricity. Yes, maybe I'd want to clean my sheets more often, but ultimately there's just Sophie, a few dust mites, and myself in the bed, so what's the big deal?

Then again, that's a pretty depressing statistic. It really *is* just Sophie, a few dust mites, and myself here—are we really making a difference? Even if sleeping naked is green, what's the point if nobody other than a silly girl in Canada is doing it?

Sadly, this cycle of cynicism has been plaguing me lately. I mean, sure, I'm convincing a few faithful blog readers every now and then to switch shampoo or turn their thermostats down, and my poor cat is getting dragged into as many eco-friendly changes as I can dream up. But as my Treehugger friend Lloyd pointed out the other day when we met up at Fresh to talk about eco-friendly building design (or rather, he talked about eco-friendly building design while I ate vegan carrot cake and tried to listen): "The greenest thing you can do with a house is fill it." Right now, my living room is looking pretty dead, my kitchen doesn't have anything cooking in it, my closets are half empty, and cobwebs are starting to form on one side of the bed. This is pathetic.

The reality is, I can do all the green things in the world, make all the difference I possibly can, but in the end it means absolutely nothing if I don't have someone next to me. Fine, maybe I'm being melodramatic, and yes, I have my friends and family, but it truly is getting tiresome putting in all this effort to help the earth, expecting to be fulfilled and then coming home, every day, to emptiness. Some people will argue that being an environmentalist and leaving a small footprint eventually leads to a sort of spiritual fulfillment, and this is accurate to the extent that buying less crap can make one feel liberated from the shackles of consumerism. Unfortunately, my spirituality tends to be rooted in the secular, or at least in the social, so when I succeed in this challenge—by, say, preparing a free-range omelette with fried yams and nonrefrigerated spinach on the side that actually tastes good—I want to share it with someone, not sit at the kitchen counter feeling proud, listening to myself chew for ten minutes. And when I fail—if, for example, too much snarkiness makes its way into a blog post and I offend people before caving in and ordering genetically modified chicken wings encased in Styrofoam, then break down and decide I just don't want to do this anymore—I want someone there. Not just my family or Meghan or my cat, but someone I love deeply, who loves me in return, who will hold me and tell me everything's going to be okay and take me to bed.

Then, I'll be sleeping naked just as often—and not just for the environmental benefits.

JUNE 14, DAY 106

Use gauze rather than Band-Aids

I got a paper cut this morning at work and the first thing I thought wasn't "Ow, crap!" Nor was it "Maybe if I suck on it for a while, that'll help." It was: "Ooh, I wonder if I could green my first aid?"

This is not normal.

And yet, whether I can make any given aspect of my life eco-friendly in some way is a question that's permanently lodged itself in the nether regions of my brain, cropping up whenever I least expect it. I'll be trying to decide whether to call a friend or send a text message, and the answer doesn't come to me based on which option is faster or cheaper—it all depends on which is greener. Which uses less energy, calling or text-messaging? Maybe I should just send an e-mail, or use Skype, or find a passenger pigeon, or a singing telegram that could arrive there via bicycle. Sometimes I'll just be sitting on the couch doing absolutely nothing and think to myself, "How could this activity be greener? There's gotta be a way . . . can I breathe less?"

It's sad, really. And now that it's gotten to the point where not even the sight of my own oozing blood will make me respond in a normal, non-green-related way, I'm starting to have some concerns. Being an environmentalist shouldn't be such a die-hard commitment, nor should it come down to such finicky questions about the carbon footprint of a Band-Aid. (Although, one really might ask why Band-Aids need to come packaged in unrecyclable wax paper, attached to disposable peel-off stickies, and be manufactured almost entirely from plastic, adhesive glue, and bleached cotton. Just saying.)

After finding a roll of gauze at work, sitting in a first-aid kit that could have been featured on the *Antiques Roadshow,* I wrote my blog post and noticed a few hours later that a commenter with the screen name keepbreathing (an anesthesiologist, perhaps?) had written: "The amount of waste we generate [in the medical world] is simply astonishing: in the course of one resuscitation, I would estimate that we could fill an entire 20-gallon trash bag with all the used and discarded plastic, syringes, wrappings, papers, and vials that get strewn about. That plus the printer in the RT office where I work spits out a non-reusable redundant second page with every single order, which can mean up to 500 pages a day being wasted. It's ridiculous."

Later that day, when I brought up the topic with my mother, a rather pro-pharmaceuticals, anti-naturopathic family doctor, she surprised me—instead of rolling her eyes and asking if I planned on attacking the modern medical system yet again with my flawed hippie logic, she pointed out that the industry could, in fact, do with a green makeover.

"There's a lot of waste in my office," she said. "Everything is disposable. It used to be that doctors would have greens—green sheets, green gowns, green caps on our heads—and it was all reused but sterile.

"Then between about 1985 and 1988, when I was in training, this changed and everything became paper—paper gowns, paper on the bench, paper sheets. I remember asking, 'Why don't we have the greens here anymore?,' and they'd say, 'Oh it's cheaper.' So everything is thrown out, and on top of that it's all wrapped in cellophane and none of it is recyclable, of course. Then there are the disposable tongue depressors, the little plastic ends on the otoscope, disposable speculums and syringes."

So did this all have to do with saving money? Even for a struggling public health care system, it seemed a little extreme.

"Well, it's not just that," she said. "I mean, people these days have to be extra careful about transmitting disease, and patients like the paper gowns and sheets because they know no one else has used them, so they're superclean. Even I've been to doctors' offices where they've given me a crumpled-up gown, and it's obviously been washed and everything, but it's sort of a bit gross—like, how washed was it?"

"Yeah, okay," I said, "but what if we could just get over this neurosis about avoiding bacteria and germs? Even some doctors are saying we shouldn't be using antibacterial hand soaps now."

"Well, the other thing is, if we went back to using cloth gowns, what that means is that when you do laundry and wash it, it has to be at really high temperatures to kill all of the viruses. Then you have to wrap the sheets inside other sheets to transport them, then steam those at a high temperature again, and then put on gloves to open them all up.

"And if you go back to using metal specula and whatnot, you've got to wash them in soapy water and clean them, then soak them in an antiviral solution, then they have to be autoclaved to make sure everything's killed off, and the nurses also have to be paid for their time to do all this, so it can just get costly and inefficient. Even from an environmental standpoint, you've got the energy required to get those hot temperatures and the antiviral solutions are incredibly toxic so you've got to get rid of that properly after you've used it. I think it probably evens out in the end. But maybe if they switched to recycled-paper gowns and sheets and didn't bother putting a waxy coating on them or dyeing them that blue color, it would at least be an improvement."

Readers, meet my mother, an environmentalist.

JUNE 19, DAY 111

Raise environmental awareness with stickers, blogs, and other media

The concept of activism provokes so much secondary embarrassment in me. I think of sandwich boards with poor grammar, unnecessary noise pollution in the form of yelling, and juvenile rhyming schemes disguised as rallying cries.

But I've discovered a form of activism I can handle: stickers. I purchased some THESE COME FROM TREES stickers from a nonprofit organization, which uses the money they receive to cover the costs of shipping and handling, with any leftovers going to the Sierra Club. Although the stickers themselves probably came from trees at some point down the manufacturing line, I feel that the results they achieve prevention-wise are worth it, and so I've started leaving them on paper towel dispensers in public bathrooms, in what might just be my first attempt at vandalism. Is it even vandalism? I'm not entirely sure, but what a rush—part of me can't believe I haven't been caught yet, that a bathroom attendant at the movie theater or a restaurant hasn't come running after me, threatening civil arrest or a lawsuit for defamation of private property. I don't have the nerve to put any of the stickers up in the bathrooms at work yet, but maybe one of these days, when the editorial board runs yet another one of its outlandish tirades about how Al Gore is the devil's spawn, I'll stick it to them.

JUNE 21, DAY 113

Cure hangovers without pills or plastic barf bags

Let's play a round of *Family Feud.* I'll be your host.

The first topic is: Most popular questions Vanessa gets asked these days . . . Most popular questions Vanessa gets asked . . . Think about it now.

Oh, forget it, I'll just tell you:

1) How's the green challenge going?
2) Have you planned the entire year?
3) How are you going to think of 365 changes?
4) What was today's change?
5) How do you keep track of them all?
6) Are you okay?
7) No really, are you okay?

I spent the first few months of this project responding to these questions with a combination of nervous laughter and slight variations of a rehearsed statement, which went something like this: "Yeah, it's crazy, I don't know, I haven't planned it all out but I'll see how far I can get, I mean I can't even remember today's change, actually, I write them the night before and they're all just blurring together now anyway, but yeah, it's okay, I'm good—seriously!"

Underlying this is usually varying degrees of panic at the fact that I still haven't thought of the next day's green change, let alone an entire year's worth.

But now that I've made it to the second quarter, the panic has subsided. While I don't have any more confidence that I can pull this whole thing off, I have come to terms with the fact that even if it dies a spluttering, shameful death, I'll still have a job and an apartment and family and friends (in fact, probably more friends because I won't have to spend every single night on my couch Googling "indoor vermiculture techniques").

However, it seems that in this twelve-month program, after the acceptance stage comes weariness. It's like some sort of hangover that's hit me before I've even gotten drunk (which I happened to do last night, by the way, and which I am greening today by sucking on bits of ginger root rather than popping Advil). In a way, it's not that surprising—I've just been greening myself too quickly, too intensely, to the point where I want to throw it all up and collapse, then wake up the next afternoon, eat a large bag of chips, chug some coffee, log onto eBay, and blow my last paycheck on some first-edition Dostoyevsky tome I'll never have time to read.

It's not a good sign that I feel like this already. In fact, I think I even look and smell like it. I was at my parents' house for dinner the other night and when I gave my mom a hello hug, she told me I stank. I explained that I was still experimenting with deodorants and that I'd been biking around the city all day. She ignored this and told me my hair looked like crap, then made a backhanded joke about how I'd never find a boyfriend until this challenge was over. This is what British parents do, by the way—they say mean things and then laugh, as if to show they're just joking, but really they're being quite serious. And my mother, of late, has been very serious about monitoring my biological clock. She doesn't care about marriage all that much, but she wants a grandchild now, and makes a point of recounting all the horror stories of her former patients—without breaking confidentiality rules, of course—who thought they were still young enough to conceive but weren't, and who are now miserable. Her message tonight was thus loud and clear: no deodorant = no babies. Anyway, this routine was pretty familiar to me, so I was able to counter in more or less the same breath that I'd rather smell bad and remain single for the rest of my life than clog my pores with aluminum, get Alzheimer's and breast cancer, and die a premature death (children of British parents, by the way, tend to be rather defensive). She rolled her eyes. I sighed heavily. We moved on and spoke of kitchen renovations.

Perhaps the best way to cope with criticism, prevent green hangovers, and stay focused on the ultimate goal of this challenge without looking like a total mess is to make like a celebrity and surround myself with an entourage. I could have Meghan as my cheerleader and nutritionist, Emma could work on my branding, Justin could be my personal trainer and smoke cigarettes beside me during morning jogs, my mother could be the on-call physician, and my dad could manage my cash flow.

Now all I needed was a dealer and a therapist.

But what might be an even better idea is finding myself a mentor—a real-life, high-achieving environmentalist. It would have to be someone other than Al Gore, someone I could call up and talk to, someone with a bit of edge. Maybe Lloyd? But then he was always so busy with Treehugger work and, besides, I'm not sure his wife would appreciate me calling him up after midnight in a fit of eco-resentment after too much syrupy Ontario wine and not enough television.

The person I usually call when I'm in such a state, when I need to vent but I want more than just Meghan's kind reassurances, is my friend Ian. He's in the same high school gang that includes Meghan, Jacob, Matt, and a few others—in fact, he's probably my best friend, if you divide shared interests and opinions by soul-baring conversation, and multiply this by fifteen years. Throughout our student days we were, as the expression goes, joined at the hip. Actually, it still gets a little frightening when we're together because we slip into this rapid-fire banter, almost a twin-speak, with slews of mental shortcuts and ultraspecific pop culture references. We can go on for hours, to the point where, if we're out together in a social situation, it's often necessary to separate us.

I've come to believe that Ian and I are most alike in the way that our earnestness is constantly at odds with our irreverence. For example, he recently left a job and a boyfriend in Montreal and returned to Toronto—which I'm thrilled about—and wanted to fill his calendar with activities. After the stress of a new gig in health policy set in, he decided to explore meditation but wanted to focus on a more practical form. Meghan—who realigns her chakras twice a day in a stylishly decorated meditation nook above her kitchen—recommended a class she'd taken last year. So Ian went.

Immediately after the first session, he met up with me over dinner to relate the hilarity that was eating a single prune for half an hour, then getting assigned mindfulness homework. If the students couldn't complete it, he said, they had to accept this fate and forgive themselves. One woman in the class harbored some sort of a prune phobia and had brought a cherry tomato instead—they all forgave her, too. Ian and I managed to go on enough tangents from here to keep ourselves laughing for hours.

No, he wouldn't be a good mentor at all, no matter what his real feelings about the environment were. He'd look at the bowlfuls of murky pasta water sitting on my counter, the unplugged fridge, and my bike stickers with slogans like MEND YOUR FUELISH WAYS! and make a smiling, derisive remark. Then I'd never be able to take myself seriously again, at least not for another 252 days. But I was running out of possibilities—why was it so hard to find a dedicated environmentalist with a self-deprecating sense of humor who could take phone calls after midnight?

Really, it's not asking much.

JUNE 23, DAY 115

Rent only hybrid cars

Last week, Mom decided to fly out to the east coast and visit Auriel, one of her oldest friends, who lives in the small town of Rose Bay on the southern shore of Nova Scotia, a place where you're guaranteed both a warm reception and a cold wind.

"Come along," Mom said. "You can see Auriel and the dogs, and walk on the beach, and eat fresh lobster and those scones you like from the Biscuit Eater."

I did like those scones.

"The house has Internet, so you can still do the blog, but then you can relax after—it'll be a nice long weekend, very salubrious, you'll see."

"Are you paying for my flight?" I asked. "Because my wallet wants to be salubrious, too."

She said fine, and so while I went over to the computer to offset the trip at TerraPass, Mom went into the kitchen and told Dad to book us a rental car at some point the next day.

"The smallest car in the lot," she reminded him.

The next day came. Dad phoned Hertz.

"The biggest car in the lot," he said.

Well, maybe he didn't say this exactly, but there was definitely some miscommunication about the difference between a compact and a subcompact car.

Long story short:

We got to Halifax. We got to Hertz. We got keys.

After filling out the paperwork, we walked over to the parking space that corresponded with the keys and said the words "Oh. My. God." Yes, in unison. Yes, as three separate sentences.

This is because we were looking at an SUV. But truly, "sport utility vehicle" is a major understatement. This was a mechanical beast. A Chrysler Pacifica beast, to be exact, in a color I believe is most commonly known as "putz white."

The two of us are both over five-foot-seven and yet when we opened the doors and tried to enter this monstrosity, we had to grab on to the opposite sides of the seats and haul ourselves up. I clambored in first and then reached over to lend my mother a hand, although she really could've used a few ropes and a carabiner.

We closed the doors. It smelled like new car.

Unfortunately, I now realize that "new car" actually translates into "off gas," which is highly neurotoxic.

It turned out the beast was actually as new as it gets. With less than a mile on the odometer, we were the first people to drive it. This felt wrong—offensively wrong. Violating, even. An environmentalist taking an SUV's virginity, or perhaps the other way around. As we got started, it felt, indeed, much less like we were driving the car than it was driving us.

"Whoa," my mother said at one point, after a brief pause in conversation during her turn behind the wheel.

"What is it?"

"This is actually starting to feel a bit scary," she said. "I haven't had my foot on the gas pedal for at least five minutes and it's keeping the same speed. And we're not going downhill."

It was true, the beast had major kick, and it was kicking Mother Nature's ass.

The sad thing is, as we realized later, Hertz actually has a Green Collection—a fleet of cars that, if not hybrids, are at least small and fuel-efficient. How did this get by us? Regardless, when we finally pulled up to Auriel's house and stepped—or rather dropped—out of the car, I turned to look around at the front yard and couldn't believe what I saw: a deer. A real deer.

Let me say, when you've been born and raised in the city, your definition of wildlife tends to be limited to raccoons in the trash, squirrels on the pavement, and the occasional rat along the subway tracks. I kept staring at the deer in front of me, and it kept staring back, no headlights required. My mom could see it now, too. Finally, it cocked its head slightly, as if to look not at us but at the sin-mobile parked smack in the middle of his dining room, still humming and crackling from the two-hour journey like it was ready for more. A second later, Auriel came out to greet us and the deer bounded off. It was enough to make me swear on the spot to never rent anything other than a Prius.

JUNE 25, DAY 117

Sell my car

As the rental car experience taught me, it's beyond hypocritical to call yourself an environmentalist and drive to work every day. It's one thing to claim hippie status and rent a hybrid, use autoshare programs, or limit driving to the weekends, but no amount of solar-paneling and green-roofing will make up for a gas guzzler in the driveway. And in my case, not even 365 eco-friendly changes are going to make up for all the driving I do. Besides, I've chosen my two green vices—not limiting myself to a vegetarian diet, and taking nine flights halfway around the world for vacation purposes—and the reality is, I don't love my car enough to add driving to that list. My precious Bugaboo will always have a place in my heart, but her financial baggage was starting to weigh me down.

What ultimately gave me the strength, however, to list a bunch of "for sale" ads on Craigslist, Facebook's marketplace, and in *AutoTrader* magazine was my newfound realization that not all big changes have big repercussions.

Take my fridge.

At first, when I unplugged it, there was fear. I'd never known a life without refrigeration and predicted my ignorance of preservatives and natural shelf lives would lead to mountains of wasted food. But after some quick e-mail consultation with Greenpa over at Little Blog in the Big Woods, who's been living happily fridgeless for thirty years regardless of his sanity, I realized how little of what I consume actually requires cold storage. Not a single condiment other than mayonnaise ever needs to be kept in the fridge, providing you use it within eight to twelve months. Fruit should never be refrigerated—it actually needs warm air to ripen. Vegetables like bell peppers, zucchini, potatoes, squash, onions, and so on don't need cold, and neither do eggs, which spend most of their life under the warmth of a hen, anyway. Carrots, I've discovered, last for a few days if they're kept submerged in water, and I've had similar success leaving bouquets of spinach and kale in vases by the window. Milk is trickier, but I mostly drink rice, soy, hemp, or almond milk these days, which last at least seventy-two hours unrefrigerated. Hummus and dips are more challenging, too, but it forces me to make my own, which usually ends up tasting better anyway. I miss cold water and chilled glasses of pinot grigio, but then there's always the option of storing a few beers in the toilet tank and switching to red wine. Butter? There's a reason the butter bell was invented—it not only keeps the stuff fresh for weeks, it keeps it at the perfect spreading consistency. Cheese? There's a reason cheesecloth was invented, too. Same goes for bread boxes. And meat? Well, okay, that's the only time I'm stumped, but all this means is that I have to buy my hamburger the day I intend on eating it, which isn't too much of a hassle as I live just a few blocks down from the Healthy Butcher.

The whole thing has been such a success that I'm half-tempted to write a cookbook or guide to fridgeless eating. It's also been one of those changes that provokes actual jaw drops in people who find out about it—no one can believe I've done such a thing until they start asking me reams of questions, most of which begin with "But what about . . ." and eventually they realize that, yes, it's true, I unplugged my fridge and lived to tell about it. I don't even have any regrets or cravings or E. coli poisoning. (I did, however, end up gorging on a carton of refrigerated vanilla yogurt at my parents' house recently, so perhaps my cravings are just being supressed.)

Meanwhile, all these little changes I've made, ones that I thought would be easy-peasy to keep up, are now haunting me. The switch from incandescents to CFLs was one, but even more irritating is that I have to save every ounce of leftover dishwater, pasta water, and other forms of graywater for my houseplants and that I have to ask cashiers every single time I buy something if they can *not* print out a receipt when inevitably the machine is wired so that it prints one out automatically, in which case I have to ask if they can recycle it and, if not, I have to do that myself or reuse it somehow.

It's these small changes that are truly getting to me, I think, because in the grand scheme of things they seem so inconsequential. But on top of this, it has to do with convenience and options—or rather, inconvenience and lack of options. Case in point: if you are trying to lose weight, you need to cut temptation off at the pass; you need to clear out all the cupboards of sugar and saturated fat, and not even go near the snack aisle of the grocery store. Similarly, in reducing my carbon footprint, I should be taking more drastic precautionary measures: if my fridge were still plugged in and cold, I'd probably give up and use it, but because it's not, this would now mean hauling the thing out, plugging it in again, rolling it back, resetting everything, and, in the end, it's not worth it.

So my hope today is that, by selling the car and cutting off the option to drive, I will be forcing myself to use alternative modes of transportation, but at the same time allowing myself all the mental and physical benefits of bike rides, long walks, people-watching, reading on the subway, fresh air, sunlight, and so on. I might still get stuck in traffic, but at least the chances are greater that I'll be in a bus and able to pass the time reading or daydreaming instead of honking at the bumper in front of me and cursing at the monopoly Céline Dion has on every radio station.

When I posted all the sale ads, I figured the majority of responses would come from *AutoTrader* or Craigslist, but instead, the most interest came from Facebook, and the best candidate was a student looking for a used but girly car she and her mom could share. We arranged to meet for a test drive, went around the block a few times, and proceeded to haggle in the most inexperienced, apologetic, and giggly way over the price. Finally, on the doorstep outside my building, as bouncy joggers, stroller-pushers, and dog-walkers whizzed by, we determined that for $11,500, my precious Bugaboo would be hers.

I took a deep breath, gave the car a final pat on the hood and a kiss on the headlight, and swallowed the remains of my emotional attachment to those thirty thousand miles before bypassing the elevator and walking back up the stairs to my apartment.

JUNE 30, DAY 122

Iron clothes only for special occasions

Before unplugging my iron, more or less permanently, I decided to indulge in one last wrinkle-pressing binge. I not only steamed the heck out of my tops, skirts, and pants but went on to tackle bed linens, pillowcases, and even my handkerchiefs. I'd been going through a lot of hankies recently, on account of a persistent cold, and was having my first truly major regret about a green change—specifically, abandoning Kleenex.

While old-fashioned cotton handkerchiefs are perfectly fine for day-to-day runniness, discreet mouth-wiping, and passionately waving goodbye to someone from a train platform, you need at least five thousand of them readily available when a cold reaches its phlegmiest stage, and that's about 4,995 more than will fit in my purse. Plus, I own only three handkerchiefs.

Well, no, that's not exactly true.

My grandma—my mom's mom—passed away last week. The funeral was in England, where everyone in my family is from, and I wasn't able to attend, so my mother brought back a few keepsakes for me: a silver nail file (my grandmother always kept her nails in perfect condition—something I envied as a girl with a biting habit, flaky cuticles, and a collection of hangnails), as well as a pair of tweezers and a couple of handkerchiefs. They were simple but delicate, white with some lace detailing around the edges. I liked that they would help me remember her not as an old lady with a valuable jewelry collection, but as a woman who knew the value of life, who was always prepared, and who respected the earth and those around her enough not to leave damp, crumpled wads of used tissue lying around. I'm not sure how she endured the snottiest days of her colds, but she made it through two world wars, so I'm sure it was the least of her concerns.

Still, whenever I try to pick up where she left off and use these beautiful handkerchiefs, I just can't bring myself to do it—the memory is too strong, and blowing my nose in them wouldn't feel right. So in the end, I'm really left with just three organic cotton ones. The cheapskate in me refuses to buy more, regardless of the fact that I'm currently expelling huge gobs of brain matter from my schnozz at least every few minutes. In dire circumstances, when I'm caught outside with no handkerchief, no long sleeves, and not even a piece of newspaper, I've been tempted to do what hockey players and some old men in my neighborhood do: simply lean over, plug one nostril, and give 'er.

That hasn't happened yet, but I'm getting desperate.

As I finished steaming the last wrinkle out of a stubbornly shriveled pair of linen pants, I could feel yet another urgent delivery from my sinuses. I glanced over to see what was left in the pile and saw my bed sheet—except to me, it didn't look like a bed sheet anymore, it looked like an enormous handkerchief, and at that moment I wanted nothing more than to gather up the whole thing, plunge my face in it, and blow.

So I did. And it felt fantastic.

Five seconds later, it felt somewhat less fantastic, realizing I'd have to clean up this patch of snot and let the thing dry again, but then I had an idea: this was one of my oldest bed sheets, one that was handed down from my parents to my sister, then to the guest room and finally to me, and it was getting kind of threadbare and yellowed in some spots (not only the spot where I'd just blown my nose). Perhaps my momentary handkerchief hallucination was in fact an inspired vision—if I took a pair of scissors to this sheet, I could probably get at least a hundred squares of solid nose-blowing material. Plus, it would mean reusing material I already had. Genius! I turned my iron off, ran downstairs to grab my scissors, and dashed back up to start snipping. Sophie looked up briefly from her napping spot on the bare mattress as if to say "What's she getting up to now?" and, subsequently, "Do I care?" then recurled herself into an unenthused ball and fell back asleep.

After a short while, I had an impressive stack of new makeshift handkerchiefs. I tried one out and it worked (how it might not have worked I have no idea, but it was still fun). Then I tried another, and another, and another. I managed to get through a good dozen of them before my mucus membranes finally gave up and dried out. It felt like ecstasy, pure lightness, like a pound of butter had just melted away from inside my head. I could hear Jimmy Cliff singing, "I can see clearly now" in the background.

Perhaps my shirts would be shabby and wrinkled from now on, but at least my sleeves won't be covered in snot.

Grandma would be proud.

july

1	Give up hair-straightening iron
2	Build a vermiculture (worm-based) compost bin
3	Shower in the dark
4	Cook and eat with the same utensils
5	No more straws
6	Cut hair short
7	Use a natural kitchen and bathroom cleaner
8	Use biodegradable pens
9	Order photos in bulk
10	No more Swiffer products
11	No more tabloid magazines
12	Stay only at eco-friendly hotels
13	Buy only locally made and sustainable clothing
14	No more canned beverages
15	No more bottled beverages or juice boxes
16	Purchase local, fair-trade flowers
17	Donate to a green cause once a month
18	Drink only organic hard liquor
19	Eat only free-range eggs
20	Eat organic dairy and rennet-free cheese
21	No joyrides on motorcycles and Jet Skis
22	No electrical forms of exercise
23	Eat food from the pot or pan I cook it in
24	Fix things rather than replace them
25	Bring my own headphones on the plane
26	Use biodegradable garbage bags
27	Volunteer at local green organizations
28	Use the greenest kitty litter tray liners
29	Switch to all-natural toilet bowl cleaner
30	Quit social smoking
31	Buy only organic cotton or bamboo bed sheets

JULY 1, DAY 123

Give up hair-straightening iron

I disagree with my mother about the aesthetic and pheremonal repercussions of switching to natural makeup and aluminum-free deodorant, but I do think there's probably a direct correlation between the frizziness of my hair and the amount of time I remain single. This is why, today, it felt like I was making another seemingly small change that would have hugely frustrating consequences—I wasn't just unplugging my straightening iron; I was unplugging my sex appeal, too.

JULY 2, DAY 124

Build a vermiculture (worm-based) compost bin

You can do it. We can help.

This is the slogan for Home Depot, the do-it-yourself mecca where one can find useful things like pressure cookers and matching towel sets as well as lots of boring stuff like plasterboard and fiberglass. Normally, if there were ever a remote chance that someone could find me in a place like this, it would be in the paint aisle by the sample swatches, daydreaming into a pastel mosaic of Martha Stewart Colors, contemplating how my bedroom walls might look in Fern Shoot green or Rolling Pin beige.

But I found myself far from the interior decorating aisles this afternoon. I was in aisle 2. That's lumber. I was also in a billowy floral-print skirt and kitten heels, aimlessly clip-clopping around and looking generally lost and confused—which was easy, because I was genuinely lost and confused.

The time had come in my green challenge to stop wasting food and start composting—vermicomposting, in fact, because while inviting five dozen worms to come live with me in a seven-hundred-square-foot apartment may sound gross, it's even more revolting having to contend with festering stir-fries, blankets of mold, and fruit-fly infestations. At least if there are worms in the mix, they'll break everything down at a faster rate, reduce the smell, and poop out natural fertilizer for my balcony herb garden.

Although Home Depot does sell a ready-made composting unit, available in their gardening department, it was a) enormous, b) made entirely out of plastic, and c) cost upward of $100. I'd searched online for other options, but it was slim pickings unless I wanted to drive all the way out to cottage country. So in the end, I decided to roll up my sleeves and make a bin myself, from scratch, using real wood.

I poked around a few nature blogs for preliminary research, then e-mailed Colin over at No Impact Man to get some tips. I figured he would probably know best, seeing as he was living in a similarly confined cosmopolitan space. His bin, as he explained to me, was essentially just a wooden box he found on the street. He had dumped some soil, shredded newspaper, and red worms into it, then gradually started adding his food scraps, keeping it covered with a lid at all times. I would just need to make sure, he explained, that the compost was ventilated and stirred around every so often. Also, I should feed it a strictly vegan diet with the exception of a few eggshells, and keep in mind that while coffee grounds, dryer lint, and cat hair are all fine, citrus must be kept to a minimum in order to balance the pH of the soil.

Sounded a bit intimidating, but still, it was mostly common sense. And other compost-savvy bloggers had mentioned that it was all about trial and error.

I decided that a simple box lined with some chicken wire would suffice, maybe with a sliding tray at the bottom to catch the worm droppings. So in between a press screening for yet another Hollywood blockbuster and an afternoon editorial meeting at the office, I stopped into Home Depot.

Standing there in my skirt and heels, with stacks of lumber towering over me, my feet getting clammy as my nose began twitching with all the dust in the air, I felt completely unskilled and out of place. What was I doing, honestly? I was a silly downtown girl in an uptown box store, floundering in a sea of orange aprons and drill bits. Suddenly, I wasn't so sure about that slogan after all—I couldn't do it, so there was no point helping.

Just then, one of the orange aprons began drifting toward me.

As it got closer, I saw that the name of the person attached to it was Bruce—a black Sharpie had been used to write BRUCE in all caps in the upper left-hand corner. I looked up and was met with a ruddy but kind face; stereotypically Canadian, it belonged to a forty-something man and was comprised of a kempt strawberry-blond beard, eyes that might actually be described as Fern Shoot green, and a wide mouth. It was a mouth that was surely about to open and direct me to the paints aisle.

"Can I help you?" it said instead. He said.

"Yes, please," I replied, repeating the mantra, "I can do it, Bruce can help, I can do it, Bruce can help," in my head.

"I need to build a compost bin," I said. He frowned. "Look, I know you sell composting units for the garden but I need something that will fit on my balcony, and I'm hoping that maybe I could try building one and you could help, but I really don't know what I'm doing or where to begin."

Phew.

Bruce's eyes darted over my right shoulder as he put his hands on his hips. Probably looking for an escape route. I started to apologize for the ultraspecific and yet completely vague project I'd just presented him with, but he quickly began nodding and asked if I had any numbers or visual aids.

I said I did, and promptly directed his attention to the diagram I'd scrawled on the back of a movie stub, as well as its corresponding dimensions, outlined on the back of my hand.

He brought out a pencil and paper.

We discussed the best type of box to build—yes, there are actually multiple ways of building a box—and examined what materials would be needed, how much it would cost, whether the wood could be acquired from the scrap pile and whether a staple gun was absolutely necessary. Eventually, I had all the stuff I needed in one shopping cart. I had it put aside, then went into work and came back later in the day with a Zipcar to take everything home. At the checkout, it came to just over $80, more than I wanted to spend but still less than all the commercial options. Plus, I now had a staple gun! The tool that keeps on giving.

I got home and lugged the planks of wood, a roll of quarter-inch mesh wire, bag of soil, two hinges, a drawer handle, electric drill, nails, varnish, staples, and staple gun upstairs. Then I unfurled one of my old quilts, laid it out on the living room floor, and sprawled the embryonic components of what would eventually become my new compost bin on top. My instincts told me to search for the IKEA guide and Allen key but neither was to be found. It was just me, and these tools, and this mess. I poured myself a warm gin martini and called my agent.

Sam, also a former colleague of my father's from when they both worked for a wireless technology company, is everything an agent should be—an ego-booster, social networker, shameless promoter, and a handyman. He likes to keep this aspect of his personality on the down-low, preferring that the only tool people see in his hands is a corkscrew, but when I explained my situation, he took pity.

"I'm staring at a pile of wood in my living room," I said. "It needs to become a compost bin by midnight. Will you come over?"

After a pause, I threw in some high-meets-low rhetoric: "It'll be fun—construction and cocktails!"

As a man with a busy social life, Sam probably wasn't thrilled about committing to this; however, as a hipster literary agent, he just couldn't resist the irony.

"I'll be there as soon as I get off work," he said.

In total, the project took only about forty-five minutes. Bruce's cutting job wasn't as precise as it could have been, so some pieces had to be crammed together forcefully. Plus, with an unattached bottom drawer that rolled out and a lid that flapped open haphazardly, it was impossible to pick the thing up and move it to the balcony without contorting all of our limbs at a series of acute angles and, worst of all, putting down our drinks.

Finally, though, it was done. We were dusty, sweaty, and splintered but had a rush of accomplishment—this is what opposable thumbs are for! I threw in the soil and walked over to the kitchen to grab the bowl of food scraps I'd been saving for the past week, then tossed them on top. Holding the lid open for a few seconds, I stared inside, half-expecting it to disintegrate before my eyes, but it just sat there. I grabbed my camera, snapped a few pictures for the blog, closed the lid, and typed up the entry.

Afterward, I went out on the balcony again, just for one last peek. It looked the same. Maybe nothing was happening now, but still—I was officially composting and it felt good.

JULY 3, DAY 125
Shower in the dark

Why have I never taken a shower in the dark before? It's the perfect way to wake up while still denying the existence of daylight and all forthcoming responsibility.

JULY 6, DAY 128
Cut hair short

YYZ–LHR–TLV–MAD–PDX–YYZ. That's what my summer vacation looks like in airport code, and I think I'm ready to tackle it. I've had to plan a few weeks' worth of green changes in advance and write up some blog posts ahead of time, get as much finished as possible at the *Post* in terms of upcoming film reviews and columns, and start packing strategically. The less weight on the plane, the less fuel that's required to propel me to all these destinations; therefore, my plan is to basically wear the same pair of pants a whole lot and bathe infrequently.

It goes without saying that there will be no indulging in complimentary hotel toiletries while I'm away, so I'll have to use my own travel-sized bottles of natural products. However, as I sat on my bedroom floor this morning, pondering the size of these tiny bottles in my suitcase, the amount of hair on my head, and the number of days on the calendar, I concluded there might be some hygiene issues come Spain—so today, I made an appointment at the salon and had my hair cut, from below my shoulders to above my chin. Fewer tresses meant less shampoo and conditioner required, not to mention less time spent lathering and rinsing in the shower. I could even use smaller towels to dry it.

Then I looked over at my bulging makeup bag. Did I really need four shades of eye shadow, two blushes, a bronzer, two eyeliner pens, three lip glosses, and concealer?

Yes, obviously.

Something else would have to go. I needed a dress for my aunt's party in England, a shawl for the conservative town of Ramallah, and a swimsuit for the hotel pool in Spain. I needed to chuck something heavy. What wouldn't I use?

Finally, I saw it—or rather, her. My longtime companion, *Anna Karenina.* I've been on page 703 of that book for about four years now; I like her too much to give up, but she's heavy and demanding. Was I really going to conquer her on this trip? Probably not. Plus, I already had another two books to get through and they were both about a third of the length and in paperback. With a sigh, then, I hauled her out by the spine and slid her back on the shelf between Tolkien and Twain. My suitcase zipped shut effortlessly—no elbow drops, cannonballs, or spinning hook kicks required—and I was officially ready to attempt my first green vacation.

JULY 11, DAY 133

No more tabloid magazines

Someone who calls herself Sense of Balance just wrote a scathing comment on Green as a Thistle today about my decision to stop buying tabloids. Criticism is bound to happen with a blog, of course, and as long as it adds to the debate, I'm all for it. Well, maybe I still get really defensive and huffy for a bit, but in the long run, objectively speaking, I realize such back-and-forth dialogue is healthy and necessary. This, however—this was verging on hate mail. Here's what she said:

"You mean, you stopped washing and turned off your fridge ages ago, but only gave up this sort of long-standing mind pollution today? Is this or isn't it over the top? Is this blog written just to be self-publicizing?

"Why not try two more vigorous initiatives:

"(1) Stop publishing this fatuous blog, and save power. If you order your priorities as you just did, nobody takes you seriously anyway.

"(2) If you really want to reduce greenhouse gas emissions as much as you can in one go, stop breathing. You'll have given up the largest human source of carbon dioxide in the environment; the only trouble is that neither burial (fatty acids leaching into the soil from the corpse) nor cremation (combustion destroys the ozone layer) is suitably Green. Perhaps you could put the dead meat into orbit, unless you're worried about the ecosystem of Mars.

"Wake up, wake up! There's life after Al Gore! Read your supermarket tabloids and have fun; just don't bother sane people with your nonsense."

So, she basically told me to kill myself.

In a rage, I contemplated deleting the comment—that's the great thing about having my own blog: I can edit other people's comments or just outright delete them, as though that dissenting viewpoint were in fact never there—but I felt, as a journalist, I should let someone else have her say, no matter how crazy, wrong, or defamatory it was, and so I left it. I'm glad I did, too, because it provoked other readers to write back in my defense.

First, there was Julie.

"While I agree with the cranky poster that today's change is a bit fluffy, you've got 365 days' worth, and darn it, you deserve the leeway! You've made some serious sacrifices (no refrigerator??? no oven???) and I admire you for this undertaking. I find this blog both entertaining and inspiring—keep up the good work!"

Thank god someone appreciated the fact that I wasn't exactly slacking off with this whole challenge and even remembered that I'd turned off my oven. Then came another.

"There's nothing wrong with fluffy changes," wrote Hellcat13. "A fluffy change is still a change. You aren't ordering your changes by priority and have never professed to do so. Every little bit helps and your choices help influence the choices of others (like me!), making an even greater impact . . . Sense of Balance should take his/her own advice by powering down the computer, which would be a blessing to the rest of us."

Oooh, diss! Take that, Sense of Balance! Or should I simply refer to her as SOB?

And wait, here was still another one, from Rhett—a video blogger who runs the site Greentime, in which he and his wife, Amy, document on camera their efforts to make a series of eco-friendly changes around the house.

"Wow. Is this Green as a Thistle's first case of a troll?" he wrote. "Amazing. I can't wait until Greentime is popular enough to have a troll . . . I find celebrity gossip to be a waste of time, too, but I fail to see a lack of seriousness just because consuming tabloids went on this long. It's still a good choice even if it was a long time coming."

I just loved this. At work, I'm always receiving angry letters about my articles, sometimes written by sane readers making justifiable complaints, other times from clinically insane grammarians who go off the deep end if there's a comma splice in paragraph four of a film review; but rarely does anyone take the time to write in about a letter about an article. I wouldn't even bother to myself. So it was refreshing and truly heartwarming to see both fellow green bloggers and devoted Thistle readers stepping up—it was the first time that catch-all phrase *online community* actually made sense and truly made me feel like I wasn't going through this alone.

JULY 13, DAY 135

Buy only locally made and sustainable clothing

Annoying? Perhaps. Flaky? Sometimes. But corrupt? I didn't think environmentalists could be that bad—not even Sense of Balance. And yet, as it turns out, some green-minded people care only about the other kind of green—cash, that is—and it's really throwing a wrench in my whole plan for today's post.

Sustainable clothes are easy enough to find when it comes to certain items—organic cotton T-shirts are everywhere, Grassroots sells lots of bamboo dresses, and Preloved has tons of repurposed thrift store sweaters—but some things are impossible to get. There's no such thing as lacy, frilly underwear that's guaranteed to be manufactured by fairly paid workers in Canada. And jeans, well, I had given up on those entirely. But then I read about UJeans, a local company that makes custom-tailored denim from sustainable material and sends it to you in an envelope made from the leftover cuttings. It sounded too good to be true. And, in fact, it was.

The website was slick and had plenty of information about the jeans, including the different styles, cuts, hems, and pockets available. There was also an in-depth selection process followed by a series of forms that customers could use to enter all their measurements, and finally the check-out page. I paid $160 in total, which was expensive, but I have witnessed my sister paying upward of $400 for a pair of jeans before, so for sustainable custom-made apparel, I thought this was a justifiable cost.

I had ordered the jeans four months ago. They never came.

I e-mailed the store owner and phoned the customer service line repeatedly, but there was no answer. I filed a complaint through PayPal—still no response. After some Google searching, I discovered that other people had been scammed by this retailer, too. Unfortunately, other than registering the grievance with PayPal and the local Better Business Bureau, there wasn't much any of us could do other than sulk and gripe.

It wasn't so much the money that bothered me, but the fact that, in my attempt to be green, I'd been taken advantage of. This would never happen at the Gap.

And yet, to be honest, the stress of a sustainable denim scam is, at the moment, far less of a burden than having to keep up with more than one hundred different green rules, both in mind and in practice, while traveling in another country. I've been in England now for about a week, and have shopped in London, partied in the Cotswolds, and castle-hopped my way across the Midlands. Now, my family has arrived in the small village of Whitburn (population: 5,253), which sits within the borders of a town called South Tyneside, just outside the city of Sunderland, which is kind of near Newcastle on the northeast coast. A little-known fact about Whitburn: the late author Lewis Carroll wrote *The Walrus and the Carpenter* here. There's a statue of him outside the local library.

Another fact about Whitburn: my grandparents live here. When I go to visit, I sleep in the little cot where my dad used to sleep when he was growing up. There's not much to do, so we usually come up for only a couple of days and spend time walking along the gray pebbled beach, which seems to wash right up into the bleak sky, until finally it's time to hit the pubs for some tea and spotted dick (that's a sponge cake with raisins and custard). My sister will occasionally convince my grandmother to get up from her motorized wheelchair and try walking for a bit—not out of any altruistic concern for Nanna's health, mind you, but because Emma is so lazy she'd rather drive than walk, no matter what the vehicle.

My granddad, on the other hand—or G-Dad, as he's taken to calling himself recently—needs no convincing. The man gets up every morning at the crack of dawn and makes his way about four miles down to the beach and back at a pace that even a professional speed-walker might find too brisk. Often, he'll go out fishing in the rowboat he keeps at a dock by the sea and return with either fresh cod or mackerel in time for dinner.

Shortly after arriving in Whitburn, we settled down in my grandparents' living room, whereupon G-Dad began telling me how thrilled he was about all the green stuff I'd been doing and how he'd been reading my blog. At first, I was surprised he even knew about it, but then I recalled that my father had set up a laptop and Internet connection at the house the year before. As it turned out, my granddad became quite the computer geek and was now using Google Earth to plot fishing coordinates. I tried, then, to explain that it had been pretty successful so far, but it was difficult being environmentally friendly when I was out of my own country, far removed from my comfort zone.

"But you know," I said, looking around, "you and Nanna seem to be living pretty green here yourselves, catching your own fish and getting by without a car."

"Ay, yes," said G-Dad. "That's right, pet. And you know what else? You'll be happy to know I've never eaten any of that white bread they sell at the grocery store, even though it's cheaper, because I know it's not as good for you, being all white and that. And I don't just go out and catch me own fish, either. I go out and get the rabbits, too, and boil 'em up. Better than chicken, they are . . . do you like rabbit, love?"

"Um, well," I said, "I haven't actually had one before, at least not for dinner. But that's great you're trying to keep a free-range diet."

"Ay, is that what they're callin' it now?"

Just then, Nanna came in from the kitchen and joined the conversation, pointing out that during the Depression and wartime, they really had no option but to eat locally. In fact, if you didn't have a backyard to grow your own vegetables, you were expected to sign up for an allotment, which is a bit like a community garden, because food was increasingly scarce. Meanwhile, clothes were usually made at home with knitting needles and sewing machines and were eventually handed down to the next generation. Most people didn't have cars, and they certainly never took vacations.

I realized, now, that whatever self-pity I might have been harboring would have to be quashed immediately. I had no right to complain about how stressful it was being green when plenty of people were getting by quite happily doing everything I was doing, and more. The carbon footprints of my grandparents combined were probably one-tenth of what mine would be, no matter how many eco-friendly changes I made.

Later on, when we'd eaten and were getting ready to leave, I thanked my grandparents for the fresh air and even fresher food, promising to keep them updated on the state of an organic tomato plant I had growing on my balcony.

"Oh, but hey," I said. "Do me a favor, G-Dad—lay off those bunny rabbits."

JULY 15, DAY 137

No more bottled beverages or juice boxes

Whenever I need a photo for a blog post, I always use either Google Images—which is best for straightforward clip art—or Flickr, a photo-sharing website with higher-quality, more artistic shots, which is also nicer because each photo says up front whether you have permission to post it on a blog and because it's easy to link back to the URL. Today, I wanted a picture of a juice box, so I logged on and entered "juice box" into the search field. Sometimes the results aren't exactly what I'm looking for; if I key in "grass," for example, I might get images of stoned teenagers instead of a front lawn. This time, I got too many pictures of children sipping from juice boxes and didn't feel comfortable posting images of other people's kids, regardless of what they were doing. So instead, I tried searching for "drink box."

What did it come up with? Hundreds of photographs of people who were either drunk and boxing, drunk and in a box, or drunk with a box on their head. I wasted at least half an hour clicking through all of them, twice, before forgetting what it was I was looking for but remembering that I really wanted a beer.

Unfortunately, because beer tends to come in bottles and cans, fulfilling this craving would mean going to the nearest watering hole, sliding up to the counter and ordering a pint, then consuming it in silence, perhaps with the muffled din of a televised football game in the background. I'm not sure I could do this on my own without feeling depressed.

But the thing is, when it comes to alcohol, I've already made a few exceptions—if a restaurant doesn't offer any organic or local wine, for example, I'm still going to order a glass of something; I'll just aim for France as the country of origin instead of, say, Australia—and because I'm definitely going to keep drinking wine, regardless of it being packaged in a Tetra-Pak or bottle, I'm thinking this get-out-of-green-jail-free card should also apply to beer.

I know this will not please any of my teetotaler readers, who will surely argue that alcohol is an unnecessary indulgence, especially during a personal challenge like this, one in which I'm hoping to improve the health of the planet while bettering my own lifestyle in the process. Yet, despite the fact that alcohol isn't essential to my survival, it does play a crucial role in maintaining my sanity—after all, if I'm going to unplug my refrigerator, cut my bed sheets into handkerchiefs, sell my car, and construct a compost bin in my living room, I'm sure as hell going to have a drink afterward.

What bothers my conscience more than occasionally breaking the green rules for the sake of booze, however, is breaking them multiple times a day for no other reason than the fact that I'm away from home. It's been a week now that I've been on vacation, and already I've had to let numerous things slide food-wise because I'm not making the decisions about which restaurants we go to. I don't always sleep naked because I've been sharing hotel rooms—and sometimes beds, too—with my sister. My extra tote bags won't fit in the purse I'm carrying, which means a few purchases have ended up in disposable bags, and I know that, if I make it to Ramallah tomorrow, I'll probably end up having to drink bottled water just in case the tap water isn't safe. This, especially, will lead to major pangs of guilt every time my need to hydrate results in a demand for more plastic; then again, I suppose if the alternative is suffering major pangs of dysentery, there's not much choice in the matter. I'm hopeful that the people of Israel and Palestine are at least somewhat familiar with the concept of recycling.

JULY 16, DAY 138

Purchase local, fair-trade flowers

After zipping around the UK visiting close relatives, distant relatives, relatives I didn't even know I had, and some relatives I almost wish I didn't have, I arrived with my sister in Israel—or rather, an Israeli-occupied Palestinian territory in the West Bank, where my friend Jacob lives.

It was 1:30 in the morning, and we had just endured El Al Airlines' infamous interrogation process (a friend of mine was once asked to sing a nursery rhyme in Hebrew before being allowed to board the plane) followed by a further line of questioning at customs in Tel Aviv and an hour-long cab ride along the wall to the Kalandia checkpoint, which opens into the city of Ramallah. I had considered the possibility of public transit, in order to avoid spewing an hour's worth of exhaust fumes into the air just for the two of us, but my mother was not keen on the idea of her daughters taking buses in the middle of the night in a country where terrorist attacks are still occurring on a semiregular basis. And, when Jacob explained that doing so would require boarding a shuttle from the airport to the American Colony hotel in Jerusalem, then walking out to the bus stop on the street, flagging one down, and riding it into Ramallah, then taking a cab from the station to his apartment because it's too far to walk and there aren't any local buses running at that hour anyway, I decided we'd better just suck up both the financial and environmental costs and take Taxi Rashid all the way.

After flashing our Canadian passports to the guard at Kalandia, we drove around the roundabout and onto another "road" (translation: meandering crevasse of broken concrete), past a "playground" (translation: a single dusty slide emanating horrifically from a clown's mouth) and weaved through a series of hills and roundabouts. The driver got lost, so for a while we had to sit and wait in the car while he got his bearings. We stared at a stray donkey; it stared back at us. I could sense my normally relaxed sister reaching for her emergency stash of Ativan and monitoring her pulse rate in the back seat. The cabbie made a phone call, and, after five minutes, another car appeared behind us. Emma, at this point, was about to flip, but I could see in the side mirror that it was Jacob behind the wheel. He got out and walked toward us, saying something forcefully to the driver in Arabic before making sure we didn't get too ripped off on the fare. Finally, by about 2:30 a.m., we were safe in his car, headed back to his apartment.

I hadn't brought any flowers as a welcoming gift, fair-trade or otherwise, but then I wasn't sure Jacob was really into bunches of roses sprinkled with baby's breath; and besides, the simple act of a woman handing a man a bouquet of anything, in public, in a Palestinian territory, might cause problems.

Hugging, of course, was also not encouraged here, which frustrated me even more because while I wasn't exactly panic-stricken sitting in that cab, I do recall that as Jacob got out of his car and walked toward us, I've never wanted to hug someone more in my life. Familiarity, in a foreign country, is bliss.

The next morning, he gave us a rundown of the apartment and, taking my green interests into consideration, pointed out that most people here had solar-heated water tanks on their roofs, used line racks for drying clothes, and—well, that was about it. The tap water was probably fine to drink but he recommended bottled stuff just in case, and while an abundance of fresh local produce could be found in town, certain things like, say, biodynamic shiraz may be harder to come by. Jacob went on to confess that he probably used more plastic bags than he should—and made a point of showing me by pulling out the bottom kitchen drawer, which was indeed bursting with a rainbow of flimsy petroleum—but otherwise lived a pretty minimalist life. He also drove to a settlement outside West Jerusalem every other week to recycle all his paper and bottles.

Wait a second, I thought—West Jerusalem? He actually crosses the wall for the sake of recycling? This is some serious dedication.

I asked him more about the whole green situation here: specifically, if there was one at all and whether he thought it was futile to try to be environmentally conscious in the midst of such intense conflict. On one hand, it would seem there are more important things at stake in this place than abiding by the waste hierarchy, but on the other hand, if so many civilians are willing to die for this land, you might think they wouldn't litter on it so much.

"People are really stupid about this," he said. "The main thing everyone fights for over here is land and water, but if you don't have proper measures to control garbage, there's going to be toxic waste everywhere. No one gets that, though."

I asked if there were any recycling depots in the Palestinian areas of the West Bank and he said no, but added that both sides were pretty negligent when it came to environmental anything. Littering, it seemed, transcended borders.

"Everyone just lives like they're in North America and consumes as much junk as possible—then you go hiking in Galilee, which is beautiful, but you see Kleenex boxes scattered everywhere and garbage caught in the trees. It's pretty disgusting."

The next day, we prepared for a typical tourist outing, but Jacob made the suggestion that, at some point between Old Jerusalem and the Dead Sea, we stop by one of the neighborhoods where I could see this limited-but-functional recycling infrastructure firsthand. My sister was less enthused—being focused on perfecting her tan at the "floaty beach" more so than doing chores—but I thought it was a great idea and made the executive decision. We gathered together all the empty plastic bottles from the apartment (no cans or glass are accepted and Jacob had just done a paper run), threw it all into the trunk of the car, and headed out.

"Okay, look for a wire cage," he said eventually, after we'd gone through the checkpoint and pulled onto a residential street, located in a pristine suburban settlement, about ten minutes later.

A wire cage? Here?

"They're not really marked or anything," he said. "I usually just drive around until I see one."

It felt like we were hunting. Hunting for recycling options.

Exciting stuff.

There was an elementary school, a few single-family homes. Nothing commercial, though, and definitely no clues as to which way a recycling cage might be.

"There's one!" he said, making a sudden left, throwing the car into reverse, and backing up. I looked around but still had no idea what he was talking about.

Finally, I saw it. He was right, it really was just a big wire cage, about six feet tall, no signs, markings, or anything to signify that you might want to throw your recyclables in it, except maybe for the fact that it was painted dark green and there was a two-foot bed of empty plastic bottles already inside. We dropped our stuff through the hole at the front while Emma took a photo for the blog, and less than a minute later we were on our way again, blaring Arabo-pop on the radio and debating whether to indulge the back-seat driver's request for a McShawarma.

I smiled to myself at the effort Jacob made on a regular basis to recycle; it's something I'd probably do, too, if I were here, despite the inconvenience. Of course, young North Americans like us have been raised with recycling as a priority. For many people, however, it's something that will get done only if it's convenient, and here, the whole nondescript-wire-cage-in-an-isolated-neighborhood-on-one-side-of-the-wall-only system doesn't really make it easy.

On top of this, the pollution that could result simply from driving around trying to find one of the darn things could very well counteract the environmental payoff anyway.

JULY 20, DAY 142

Eat organic dairy and rennet-free cheese

Milk is a very divisive issue.

There are vegans who refuse to consume dairy because they believe the cows are treated poorly. There are nutritionists like Meghan who avoid it at all costs because there might be trace elements of hormones or pus in milk products and our colons don't particularly like digesting milk, especially when the lactase enzyme has been slacking off. Then there are people like my friend Matt, who lives in Paris and would probably suffer a major panic attack if he were told he couldn't have any six-year-old Camembert with his Beaujolais. There are also the dairy farmers, who of course feel passionately that drinking milk is natural. And let's be honest, what would this world be without ice cream, really?

However, when my friend Kate, who recently finished culinary school in London and eats pretty much anything and everything without putting up a fuss, told me that she insists upon having organic dairy in her kitchen, I knew something was up. The more I researched, the more I discovered that regardless of whether it was right or wrong to consume milk products—either ethically, nutritionally, or environmentally—it was definitely important that if one did choose dairy, it be organic.

After spending more time than anyone should ever spend surfing the Dairy Farmers of Canada website—teat dip is only the beginning—I ended up at the Wikipedia entry for rennet. The last bar of organic cheddar cheese I'd purchased had advertised itself as being rennet-free, and I wasn't sure why I should care.

Well, let me just say, learning about rennet was nearly enough to put me off dairy altogether.

In the simplest of terms, rennet is a collection of naturally occurring enzymes produced in every mammal's stomach that helps to digest milk. It contains protease, which coagulates the milk and separates it into solids (curds) and liquid (whey). This, in itself, is hardly disturbing. However, in order to make cheese, a farmer must initiate this process before it gets to our stomachs, so rennet is extracted from the stomach of a cow, goat, sheep, etc., in advance.

The traditional method of making rennet goes like this:

1) Clean and dry one calf stomach
2) Slice into small chunks and submerge in saltwater
3) Add vinegar or wine to lower the pH
4) Let sit overnight
5) Strain and filter

The modern method, which is interestingly no less revolting, goes something like this:

1) Clean, dry, and deep-freeze one calf stomach
2) Mill and submerge in enzyme-extracting solution
3) Add acid
4) Filter until concentrated

Hungry yet?

Apparently there are vegetarian-friendly alternatives to rennet. As the Wikipedia entry explained, "certain plants have just as effective coagulating properties, such as fig tree bark, nettles, mallow and thistles." Hey, thistles! Right on. "Enzymes from thistles, or cynara, are used in some traditional cheese production in the Mediterranean," it said. Wow, maybe I should have called my blog Green as a Natural Coagulating Agent.

But this was only a brief intermission from the horror. The next heading on the page was "Genetically Modified Rennet" and included this paragraph:

"With the development of genetic engineering, it suddenly became possible to use calf genes to modify some bacteria, fungus or yeast to make them produce chymosin. Chymosin produced by genetically modified organisms was the first artificially produced enzyme to be registered and allowed by the FDA in the USA. In 1999, about 60% of U.S. hard cheese was made with genetically engineered chymosin. One example of a commercially available genetically engineered rennet is Chymax, created by Pfizer."

Great, Pfizer's found a way into my yogurt.

Well, not anymore. From now on—and, I truly believe, for the rest of my dairy-eating life—I'm enforcing a strict organic-only, rennet-free rule.

Fortunately, in Ramallah, it's easy enough to avoid dairy. Jacob would know—he's lactose-intolerant. But the type of dairy that's most often available here is sheep or goat milk, which is produced in as old-fashioned a way as possible, and is supposedly easier to digest than cow's milk on account of the higher levels of fatty acids, the softer protein curd that forms once it's in the stomach, fewer allergens, and less lactose.

"You know," Jacob said, as we drove past a herd of goats one afternoon on our way to Jerusalem, "the Palestinian Livestock Cooperatives Union is one of our partners at Souktel, so you can think of me every time you eat Palestinian goat cheese . . . which I'm sure must be a regular occurrence at your apartment in downtown Toronto," he added.

"Wait a sec," I said. "Remind me how text-messaging benefits a goat farmer? Do the farmers send out messages to their friends, like, 'Goat milk 4evR! TTYL'?"

"Yeah, LOL! No, silly. It just helps the union keep track of all the livestock, and then the farmers can schedule vet appointments or book training sessions or whatever, but they can do it all using their cell phones. No one has Internet or landlines here, really."

"Ohhh," I said. "Pretty smart idea. Did you go to Harvard or something?"

"Very funny."

"So hey, do any of these goat and sheep farmers make organic cheese?" I asked. "I guess it might be difficult to get properly certified when you're behind a wall."

"I'm not sure," he said. "It might be accidentally organic."

"I think you should send them all a text message and ask," I said. "Oh, and while you're at it, find out if they use rennet, too. That stuff really sucks."

"What's rennet?" asked a voice from the back. My sister.

"You don't want to know," I said. "Just don't order any more Big Macs when you go to McDonalds."

"Yeah right," she said. "Like that's going to happen."

JULY 22, DAY 144

No electrical forms of exercise

On July 19, at 5:30 a.m., Jacob bid us farewell with groggy cheek kisses and promises to keep in touch. Emma and I then made our way back to the airport in Tel Aviv, where we split up so that she could go back home to Toronto while I flew off to Spain. My plan was to spend a week helping a group of Spaniards improve their English in the small town of Ávila, a few hours west of Madrid. There were about a dozen Anglos and a dozen Spaniards, and the routine for most days was: breakfast and casual conversation; three structured speaking sessions, each one-to-one, and an hour long; lunch and more conversation; another couple of one-to-ones; a group activity; dinner and even more conversation; free time.

The schedule involved a lot of walking and a lot of talking but it was a sweet deal—at least for the English speakers, who volunteered their time in exchange for a week of free food and booze, complimentary transportation to and from Madrid, and accommodations at a stunning five-star villa set among the rolling sepia hills of Spain's countryside.

It also meant my pledge not to use any electrical forms of exercise just got easier. I was taking walks into town, by the river and across the fields; swimming at least once a day and dancing at night. I was getting a strong dose of vitamin D while using the most natural sunscreen I could find—unfortunately, it turns out the mineral-based stuff I originally bought goes on gray and doesn't absorb, so I look like a corpse until I get burned, then I just look stupid. I switched to Neutrogena's SPF 30 for Sensitive Skin, a product that the Environmental Working Group green-lighted, and it was much better. I was also getting tons of natural exercise and eating predominantly local food, which wasn't technically organic, but all the farms we drove by on the way to the hotel had cows grazing outside and pigs mucking about in their pens. This struck me about Europe: a lot of people here may smoke and drink to excess, eat tons of meat, and not recycle as many different types of plastic as we do in North America, but fundamentally, they are often more sustainable. They drive smaller, more efficient cars, their agriculture practices are far more wholesome and natural, there are barely any big-box stores or strip malls, people eat home-cooked meals more often than junk food or take-out, and in general the outlook on life is simpler—eat well, drink well, love, and be happy.

So while I was at this program, I felt it was easy enough to live the green life and didn't put up a fuss about whether the coffee was fair-trade or what species of fish we were eating or what laundry detergent the hotel staff used. And I most certainly didn't stop to question whether the sangria was local—that would just be rude.

On the last night, however, we didn't drink sangria or regular wine or even this bizarre drink all the younger girls were obsessed with called Calimocho, a mix of red wine and Coke. No, we drank what's called Queimada (which translates, simply, as "burnt"). Much like absinthe or, perhaps more accurately, paint thinner, it's a mix of Orujo—a 100-proof grappa-type solution with over 50 percent alcohol—poured into a big cauldron with sugar, citrus rinds, and coffee beans. The person preparing it then lights it on fire and stirs it around with a big ladle while reading an ancient Galician incantation. Finally, it's poured out to everyone gathered; the moment you sip it, your lips suddenly feel very hot and very sweet (needless to say, at least six of the Anglos and Spaniards hooked up that night, whether on the dance floor or in their hotel rooms).

A bunch of us, however, still had some semblance of sobriety at 2 a.m. and managed to collectively stumble over to Simon's room. Simon was a young British guy who was a little too into South Africa and a lot too into golf. He'd brought his laptop with him and a bunch of music, so we all piled onto his bed and continued talking, laughing, singing, then watching TV on mute and fiddling with the song list. I was sitting next to Javier, a lawyer from Madrid in his late forties, with whom I'd had a few memorable one-to-ones. We'd attempted to discuss—in either broken or awkwardly overenunciated English—everything from whether Quebec will separate to what it means when you're "egging someone on." At one point, we ended up debating the best way to seduce a woman and whether this depended on other factors like geography, context, and age. From that conversation on, Javier had been looking at me differently. I'm not sure what it was specifically that I said, but he kept glancing in my direction with narrowed eyes and head cocked to one side. More than once he had told me how fascinating it was to him that I was "very beautiful, but with smart."

When it hit 3:15 a.m., most of us decided to go to bed, or at least get off Simon's bed. Everyone walked out of the room—including Simon, who wanted a cigarette—and just as I was about to step out the door, Javier pulled me back and kissed me.

I say kissed, but it was more like smothered and slobbered. He just went for it, tongue and all, grabbing my waist and feeling up my back as though he were prepared to start undoing my bra, and more. I managed to pry away from his grip, tell him, "No, no, no," which I'm pretty sure in Spanish also means "No, no, no," and leave. But Javier followed me out of the room, keeping pace right behind me. I turned, finally, to go up a set of stairs, then stopped, held up my hand, and said good night as emphatically as possible.

His response: "Tell me, what is your room? Which number?" This was getting ridiculous. I tried, now with more hand gestures, to express that nothing was going to happen because he had a wife and we were both drunk (and besides that, I was totally grossed out by the prospect of even contemplating getting it on with someone nearly my dad's age).

His response: "Where is number your room?"

So I told him, "2154."

My room number was actually 2153, but there was no way in Queimada I was dealing with Javier knocking at my door anytime soon. Finally, I turned and ran up the stairs while he walked away. I got to my room, undressed, and got ready for bed—putting on a set of pajamas just in case—brushed my teeth and washed my makeup off, then turned out the lights and started to doze off.

At about 3:45 a.m., I heard knocking.

But it was next door.

I held my breath—fortunately, I knew the two girls who were sharing the room beside me and could at least explain the situation the next day, but still, this was bad.

He knocked again. No answer. Thank god.

Then I could hear the footsteps coming back in my direction. I winced, cringed, and grimaced all at once, but to my relief they kept moving on toward the other end of the hall. I was safe. No more Javier. I couldn't help but replay the sloppy grope-fest in my head as I drifted off again, and kept shuddering—something about it seemed nasty in the same way that littering, factory-farmed pork, and putz-white Chrysler Pacificas were nasty. Although I wasn't exactly sure how the whole episode could be classified as un-green on any technical grounds, it just was.

Perhaps if I made a point of washing my mouth out with vinegar and baking soda the next day, I'd be able to repress the memory.

JULY 27, DAY 149

Volunteer at local green organizations

My fourth flight in less than three weeks deposited me back home in Toronto, where I'd have a bit of time to un-jetlag, do laundry, and prepare for this sustainable cycling trip. Catching up on e-mail, I noticed there was a meeting being held tonight for the Toronto Environmental Volunteers (TEV)—I wasn't quite sure what these people did, but seeing as I was from Toronto, cared about the environment, and was looking to volunteer, I decided to check it out.

Walking into the conference room at Metro Hall, I expected to find a room full of eager hippies cradling acoustic guitars and seedlings, collectively preparing for their next eco-crusade, whatever it might be. Instead, I saw the most average-looking crowd ever—you could have just taken the line at the Department of Motor Vehicles, transplanted it to this airless room, stuck a complimentary oatmeal-raisin cookie into everyone's hands, maybe some dejection into their souls, and it'd be the same thing. All right, so, fine, there was one enthusiastic guy in a Tilley hat and a smiling middle-aged woman decked out in a batik vest, dangly artisan earrings, and high-waisted khakis, but still—bo-ring.

Because this was a training session, we had to watch a PowerPoint presentation of what TEV was all about, the types of activities we'd be involved in, and so on. This was given by Janet, the smiling woman who turned out to be the head cocoordinator. No one in the room looked even remotely interested and I wondered why not. I mean, the slide show was hardly riveting material, but weren't we all gathered here to get pumped and make the world a better place? It's not as if any of these people have to be here.

Or did they? As it turned out, they did. Janet stopped the presentation at the halfway point to let people get more coffee (in ceramic mugs—check) or juice (in disposable bottles and from concentrate—uncheck), then said something about how she'd sign everyone's log books before moving on. I asked the guy in front of me what she was talking about. Turns out, the TEV is a popular choice for welfare recipients, ex-cons, and high school kids doing their mandatory volunteering hours. All these people were really looking for was a signature next to the day and time in their book, then they could leave. At least half the class was gone before the second half of the show began.

I let out a sigh and took a swig from my Sigg bottle—I'd decided in the end to get a stainless steel water bottle in place of my old plastic Nalgene when the federal government declared bisphenol-A a legitimate toxin.

With the handful that now remained, a somewhat younger but still bland bunch, Janet persevered, asking us if we could tell her how much waste is diverted from landfills each year thanks to our curbside recycling, or why the city of Toronto enforces a do-not- feed-the-birds policy at Sunnyside Beach. The answer to this second question came from an older bald man behind me who said, simply, "Crap." Only upon further prompting from Janet did he choose to elaborate: "They crap. In the water. Canada geese, worst of the lot. Drop 1.5 pounds of crap every day. One. Point. Five."

And, moving on.

Next came the slides about environment days, which are held on a regular basis during spring, summer, and fall in various neighborhoods throughout the city—it's here that residents can drop off used paint cans, dead batteries, and other hazardous waste, as well as ask volunteers questions about what can and can't be recycled. There's also a big pile of compost made from lawn clippings, leaves, and other mulchable stuff that the city collects and then leaves in an anaerobic digester for a year, which finally gets poured out in a big pile so that anyone who wants it for their garden can bring a shovel and go nuts. Sounded a little odd to me, but Janet insisted it was hugely popular and a "must-see."

She clicked through another few environment day slides and

eventually stopped at a photo of two girls hugging something. Or was it someone? Oh dear, it couldn't be.

But yes, yes it was. This was an official TEV mascot. I kept staring and squinting at the slide as it was difficult to see what exactly this creature was—from where I was sitting, it looked to be a human form with fuzzy yellow limbs, bright red underwear, and, in place of a head, an enormous droplet of water.

"This is Squirty," said Janet.

Okay, no, she has got to be kidding, I thought, wishing desperately that Ian were sitting next to me to witness this.

My hand shot up.

"Yes?" Janet said, still smiling.

"Um . . . Hi. Hello," I replied, introducing myself. I then brought up my high-pitched, superpolite voice and asked: "I'm just wondering—what exactly *is* Squirty?"

"He's our mascot," she said.

"Oh, yes, I see that," I said. "But, uh, I'm afraid I don't really get what he's supposed to be . . . Like, is he a leaky faucet? Or acid rain? Or is he just a random drop of water from somewhere?"

Janet said she didn't think Squirty was supposed to be anything in particular. He was more of a conceptual mascot.

"However," she added, "Squirty isn't the only mascot we have."

Oh god, there were others?

"If we have multiple volunteers who want to dress up, they can use these costumes—our secondary mascots."

She clicked on the next slide.

The secondary mascots were nameless, made out of foam, and more straightforward in their imagery. The one on the left was a life-size green bin, your standard residential organic waste unit, with a smaller kitchen container perched on top; the one on the right was a stack of newspapers. That's it. A stack of paper. They looked so pathetic in Janet's slide that part of me wanted to burst out laughing but another part of me felt that they were actually so sad they should be incorporated into a Gus Van Sant film. I decided to make up my own names for them, to at least give them some dignity.

The names were: Stinky the Green Bin, and Stacky the Pile of Newspapers. Stinky and Stacky for short, of course. I kept these to myself.

The training session came to an end, and as people began collecting their coats, I walked over to sign up for the regular e-newsletter as well as any upcoming volunteer duties. Squirty, Stinky, and Stacky may not be a part of anything I end up doing, but there wasn't any reason to give up on TEV just yet. As lackluster as the past couple of hours were, there was more here. Something deeper was going on behind Janet's batik façade, and I wanted to know what that was—what was motivating her to get up and do this every day? In any case, whatever I ended up doing with TEV would surely be fulfilling in some way, if not on a social level, then at least perhaps on a personal or moral level; and one thing was for sure—I wouldn't have any mindfulness homework, nor any obligation to forgive myself.

Still, the moment I stepped outside, I reached for my phone and called Ian.

JULY 31, DAY 153

Buy only organic cotton or bamboo bed sheets

These organic cotton sheets are way nicer than the red polyester ones I bought for myself back in college. And yet, no one but Sophie can appreciate them. Also, her way of showing appreciation involves leaving clumps of fur and dander everywhere. Sexy.

august

1	Enforce a shoes-off-at-the-door policy
2	Use only handheld fans
3	No more new leather
4	Ensure house sitters abide by green rules
5	Buy only recycled or eco-friendly jewelry
6	Use only cold water for laundry
7	Buy all-natural cat treats
8	Take lukewarm showers
9	Send only recycled greeting cards
10	Print everything double-sided
11	No more Wite-Out
12	Buy spices in bulk
13	Keep a brick or plastic bottle in the toilet tank
14	Buy only ethically made, sustainable footwear
15	Use a manual tire pump
16	Use biodegradable packing materials for shipping
17	Skip gown at doctor's office
18	Donate used clothing to thrift stores
19	No more staples
20	Use wind-up, LED flashlights
21	Get houseplants
22	Wear only natural perfume
23	Get massages by people, not chairs or electronic gadgets
24	Choose an electronic rather than paper ticket
25	Use only electronic press kits
26	Eat only organic tubers and root vegetables
27	Online dating with GreenSingles.com
28	Use cloth napkins instead of paper ones
29	Make jam and preserves
30	Buy only used sports equipment
31	Use all-natural face cream

AUGUST 4, DAY 157

Ensure house sitters abide by green rules

Justin had assured me that he enjoyed housesitting, that it wouldn't be any problem whatsoever to look after my apartment while I was on vacation. But I wasn't so sure. This wasn't just any house, after all—it was a green house. Instead of Kraft peanut butter in the cupboard, there was organic almond butter; instead of cold beer, there was warm Ontario wine; and instead of a regular trash can under the sink, there was a multi-hued collection of recycling boxes and a homemade compost bin on the balcony, which just so happened to be in need of worms.

But my former colleague from the *Post* insisted he wanted to get away from his roommates for a while, so I had thanked him profusely, given him a key, and left a detailed note—or rather a three-page, single-spaced Word document—explaining how best to survive my greenness, which I left on the kitchen counter. Miraculously, when I returned home from Spain, I found a spic-and-span apartment exactly as I'd left it. Justin explained in his own note that he'd turned the fridge on while I was away but it was unplugged again now; he had picked up my worms, which should be making themselves comfortable in the compost; he accidentally broke my French press but got me a new one; and he even managed to bond with Sophie, who usually doesn't take well to guests. Amazing, I thought. But then, I should have expected as much from a former copy editor—you need to be a perfectionist to do that job, and being one myself, I have to say there was something very satisfying about having two obsessive-compulsive personalities come together in such a symbiotic (and clean) way.

A sigh of relief, another note to Justin, and a couple of loads of laundry later, it was back to the Toronto airport, where I was flying to Portland via Philadelphia with not much time to change terminals. At least the gate for the second leg of the trip was easy to find—thanks not so much to any strategic architectural planning, mind you, but because it was totally crowded with hippies. They were all in ethnic tunics, perched on REI backpacks, picking at the frayed bits on their Tevas, and looking somewhere between unimpressed and hungry. I wondered whether, if a new airline were created just for this demographic—say, Hippie Air, where all flights were carbon-offset and passengers got both a vegan meal and a complimentary issue of *Mother Jones*—they would look more enthused. My gut hypothesis: probably not.

I arrived in Portland after midnight. Sadly, my luggage did not. I dragged my annoyed heels over to the baggage service counter, where there was a long line, but my spirits were immediately revived upon remembering this was America, where customer service actually meant customers got served and employees were either genuinely friendly or very good at pretending to be. I felt especially bad for these people tonight, too, having to deal with irate travelers for eight hours straight on the late shift. So when they handed me a claim ticket, apologized, said my suitcase would arrive at the hotel the next morning, and gave me a plastic pouch full of what they referred to as "emergency toiletries"—a mini-toothbrush, toothpaste, shaving cream, razor, comb, deodorant, and soap—I smiled and accepted it. But wow, I thought later, what a waste of petroleum, and in Oregon no less, one of the greenest states there is. Don't these guys have laws against things like idling, pesticides, and plastic emergency toiletry pouches?

The next morning, I woke up early. Really early. For a moment, I was able to appreciate the novelty of being able to literally roll out of bed and walk out the door (this was due to my having slept in the only clothes I had with me—sleeping naked was not happening at the Holiday Inn—and not being able to use most of the preservative-filled toiletries in the emergency pouch), but, as with most novelties, the enthusiasm quickly wore off. The orientation meeting for the bike trip was at 7 a.m. and it was already 6:30. There wasn't a single bus stop or train station anywhere near the hotel, which was practically in the suburbs, so I hopped into a cab and made my way over, figuring there would be some coffee and pastries to shove in my mouth in between the waiver signings and name games.

The meeting room was dark and sparse, except for a few seizuring fluorescent lights and a handful of inspirational posters tacked to the wall. A few people were there, sitting silently. I took a seat in one of the plastic chairs and looked around. No coffee.

"Why no coffee?" I asked in a whisper to one of the girls there, who had shiny hair and a pert, tanned nose, and who was generally very wholesome and peachy-looking—appropriately, as it turned out she was from Georgia. She said she wasn't sure, but guessed it had something to do with the fact that the tour served vegan food and maybe these people didn't like altering the natural state of their body with stimulants like caffeine.

Fuck. Okay. But why no pastries, I wondered? Oh right, they have butter—butter that comes from milk, which comes from cows, who may have been treated with antibiotics or overworked or separated from their calves too early, and besides which produce way too much methane to sustain a healthy planet.

Or something like that.

So what *was* available to eat during this six-hour orientation? I looked over at the table by the door: dry bagels, jam, and water. It reminded me of the bagel-and-juice tents at the Sporting Life run, except now, instead of having a 10K behind me, I had half of Oregon in front of me.

This was going to be sooooo much fun.

The session finally got under way. It was led by Mark, a lithe, dozy, and wide-eyed slow-talker, also with that impeccable American tan, but with a less impeccable haircut. He looked to be in his early thirties, wore a Hawaiian shirt with bright orange shorts and the requisite Tevas, and raised his eyebrows a lot, as though confused, curious, or both. Mark admitted to being an ovo-pesco vegan, meaning he didn't eat meat or dairy but chose to eat eggs and fish, which meant, at least to me, that he wasn't much of a vegan at all. This was reassuring. At various points in the presentation, he also revealed that he had a Ph.D. in neuroscience (approval rating: up), lived on his own in a mango forest for a year studying moon patterns (approval rating: down), and was a registered massage therapist (approval rating: up . . . way, way up). In fact, if there were ever a real-life mascot for Oregon, I thought—an equivalent of Squirty, perhaps, without the costume—Mark would be it.

During the second half of the orientation, we looked at slides depicting photos from previous rides as well as a number of local, sustainable initiatives, which we were free to check out at the beginning and end of our stay here in Portland. Mark told us all about the organic restaurants in the city that converted their used vegetable oil into biodiesel, about the pedestrian- and bike-friendly infrastructure, about the green-minded eco-schools, and finally about a neighborhood right around the corner whose residents decided to come together and paint a big colorful sunflower in the middle of a major intersection in order to slow down traffic. They also installed a communal message board and sharing center on the corner.

"And then they gave out free kittens and rainbows to everybody!" I quipped in a sarcastic, hushed tone to the guy sitting next to me.

"What?" he said.

Oh, never mind.

It was somewhere around this point when the culture shock set in. Going from Ramallah to Spain to Portland was about as jarring as it gets in terms of societal norms and priorities. What was most frustrating, however, was that I couldn't justify my annoyance at this situation, at least not in any logical terms. Why should I be irritated at a group of people who simply want to ride their bikes and learn about sustainable farming practices and eat healthy food that doesn't come from animals? There's nothing wrong with this. In fact, it's specifically why I signed up for this tour in the first place.

Halfway into the orientation, Mark took us outside to play a game in which he asked a series of questions like "If you know the difference between active solar and passive solar, take a step forward"; "If you've been on a multiple-day cycling trip before, take a step forward"; and "If you eat organic food, take a step forward," in order that we might get to know one another a bit better and figure out where on the green continuum we currently resided.

To my utter surprise, I probably took more steps forward than anyone else there. Clearly, then, I had no reason to dislike these people—if only because I was pretty much one of them.

Thinking back to a conversation I had with Ian recently about the ethos of the modern-day green movement, I recalled one of his comments about how, back in the sixties, hippies went all out—burning bras, getting high, tossing beer bottles or bricks into the toilet tank to save water. But today's hippies don't have the same edge: they'd never burn a bra because that would be a waste and pollute the air; they'd never drop acid or smoke anything other than organic weed because that would be putting unnecessary toxins in their bodies, which by the way are sacred; and they'd rather install a proper dual-flush toilet than use a brick as the latter could corrode the plumbing and is probably better off donated to Habitat for Humanity anyway.

Yes, this was it: the values these modern vegan cyclists seemed to maintain were fine, but their tendency toward restraint and guardedness, not to mention the fact that they take everything, including themselves, way too seriously, was getting on my nerves.

However, I'm generalizing, of course. Sure enough, when we went around the group one by one, explaining why we signed up for the tour and what we hoped to take away from it, there were some responses that didn't fit this hippie-vegan-cyclist mold. A teenage girl from Portland, for example, simply stared at the floor and said, "I just want to ride my bike." One of the older women admitted she was mostly there to figure out what this quinoa fuss was all about. Then a guy named Chris raised his hand. He was from Texas (cue gasping) and worked for Hallmark, and said he was hoping to meet some other environmentally minded people here in order to bring back some inspiration and see whether he might be able to green the mother of all greeting card companies.

Well, I thought that was pretty cool.

I also thought he was pretty cute—though sadly, I later realized, pretty married and pretty trying to have a baby.

By the end of the orientation, I knew Chris a bit better, as well as Dave, the guy who didn't get my kittens-and-rainbows comment. Dave was a thirty-something, six-foot-lanky lifetime vegan who had just munched his way through an entire backpack full of raw green peppers, carrots, broccoli, and zucchini and still looked malnourished. But I liked that he'd made a point of declining the dry bagels, and what he lacked in ironic sensibility he made up for in kindness. He was one of those people you feel naturally inclined to open up to, someone who has the rare ability to be opinionated and confident without offending a single soul.

I decided these two were probably the most normal people on the ride and so made some preliminary efforts to latch on to them. We went for a walk together that evening—Chris was into photography and wanted to get some artsy shots of Portland—and decided to grab a last supper at a nearby veggie-friendly sushi restaurant before heading out early the next morning. It would be my last glass of wine for the week, which did not make me happy. I'd thought about watering down some merlot and pouring it into my water pack (it would look like fruit-punch Gatorade, I thought); however, in the end I figured this would not only taste revolting, it would kick me straight up to the podium of an Alcoholics Anonymous meeting. I ordered another glass and tried to focus on what Dave was saying—what *was* he saying, anyway? Oh no, he was complaining to the waitress. She had brought him a miso soup possibly made with fish broth.

AUGUST 8, DAY 161

Take lukewarm showers

I love scalding hot showers, the kind where it hurts real good, where your skin wants to scream but you feel warmed right through to your spinal cord and not just clean but disinfected. However, this morning, I had the best shower of my entire life—and it was lukewarm. I was at Sunbow Farm in Corvallis, Oregon, where the owner, Harry, had installed an outdoor solar-heated shower for his workers. It had been cloudy the day before and a few people on the trip had already used it, so the water wasn't about to get piping hot. My fellow trip mates warned me of this, but my grime was getting to an uncivilized point—I looked as though I'd escaped from the Neanderthal exhibit at the local museum—and it was time for a thorough scrub. One of the girls led me over to where the shower was and explained the routine, then left me to my own devices.

I closed the burlap curtain, which went up to only about shoulder height but covered all the important bits down below, and eventually got the water going, positioning myself tentatively near the stream. Suddenly, I turned around and nearly lost my breath—not because the water was cold, but because I was looking out across vast, rustling fields of amaranth, quinoa, and barley as the sky turned an overwhelming blue, the white clouds burst like popcorn, and the breeze brought the sounds of a distant breakfast to my ears. It was total perfection. I decided, right then and there, that warm was the new hot—every shower I took from now on would be precisely this temperature.

Ah, Sunbow. Halfway through the bike trip, it was the first time I felt genuinely happy to be doing this. The previous days all had their various perks—bucolic scenery, rigorous cycling, even a special visit from an organic dairy farmer—but I had yet to make a real connection with anyone there. Day two, especially, was a flop. In fact, it gave me nightmares. We were staying overnight in Salem (fitting, in retrospect), at a place called Pringle Creek, a planned community in its final stages of being greened up the wazoo. Everything is LEED-certified, all houses have solar panels, wood is all reclaimed or approved by the forestry stewardship, all insulation is made from recycled materials, the construction equipment had been powered with biodiesel, and an organic community garden was in the works. But it was still new and, like most suburban developments, had lawns resembling golf courses and slithering cul-de-sacs everywhere. What the developers were most proud of, however, were the roads, which were made of porous asphalt. James, the tour guide, explained that when it rained, the streets absorbed 90 percent of the rainwater, channeling it underground and directing it back into the surrounding rivers and streams. Part of me thought this was neat and wanted to try pouring fresh-squeezed orange juice on it to see if the pulp would get strained, but another part of me was completely freaked out and wanted to avoid stepping on it entirely, just in case it sucked me underground, too.

Later that night, as we were setting up the tents on one of the lawns, I very nearly stepped right onto a corpse. Not a human corpse, mind you, but the dead, decomposing remains of what looked like a fetal groundhog. A few seconds later, my tent mate found more of them, covered in flies. I freaked out and started flailing my hands about my face while she got a shovel and went about scooping them up matter-of-factly, depositing them in some bushes across the road. The following morning at breakfast, three people talked about how they had nightmares and couldn't sleep. This place may be green, I thought, but like its Massachusetts namesake, it's still cursed.

Harry's farm, however, was a different story. Sunbow felt good. It felt safe. My tent was tucked in between two plerry trees—no one could figure out if they were plum or cherry so the consensus was it was some hybrid of the two fruits—and all our meals were made from organic produce and grains harvested within a hundred-meter radius. Harry knew absolutely everything there was to know about sustainable farming, permaculture, and the organic certification process, but he spoke plainly about it all so it was easy to understand and ask questions. I never thought I could be so interested in subjects like the genetic makeup of organic versus commercial tubers, but I was—I really was.

AUGUST 12, DAY 165

Buy spices in bulk

Today, I pledged to buy all my spices in bulk, but I think I might promise to eat them in bulk, too, because as much as I've grown to like all this yerba matte tea, Swiss chard, and tempeh, the reality is that vegan food can be unbearably bland.

Even the vegans here will back me up on this. Emmanuel, one of the guys my age on the trip, quickly became known for his salad dressings and homemade sauces—he has a gift for blending flavors and it came in very handy on the nights when our ingredients for dinner were limited to three forms of starch and nutritional yeast flakes, one of the only available sources of vitamin B_{12} that's not animal-derived. Most vegans will tell you it tastes a lot like butter, which I think is a fairly ironic descriptor for obvious reasons.

Spices, however, can be tricky, especially for those vegans who choose to restrict their diet based on environmental factors. Things like basil, thyme, and coriander can all be grown locally, but garlic and ginger usually come from China, cinnamon from Sri Lanka, cumin from the Middle East, vanilla and curry powder from India, and so on, which means the more flavorful your meal, the bigger carbon cost it may have.

I asked Emmanuel how he felt about this, if it mattered at all to him whether his culinary choices might mean a chicken is saved but a few thousand tons of CO_2 are coughed into the air by cargo planes and delivery trucks.

"You know, I feel anxious when I'm in the grocery line and realize I don't have my canvas bags, or if I'm out somewhere and can't find any recycling bins," he said. "There's a constant awareness of how much packaging I throw away, how many pieces of toilet paper I use and whatever, but honestly, in the context of all that I do to help the environment, I don't really consider my use of spices to be that contentious."

"But what if your demand for cinnamon on that bowl of vegan

oatmeal meant a whole extra case had to be shipped from the other side of the world?"

"Yeah, but I don't think it would," he said. "And besides, it's not like I use gobs and gobs of the stuff. Like, really, onion, garlic, and basil are my staples, and other than the occasional Indian stuff, I stick to soy sauce, olive oil, and rice wine vinegar, and I get most of my vegetables locally. So if my few dashes of cinnamon one morning come from Sri Lanka, I'll still be able to sleep at night."

Well, maybe Emmanuel would be able to sleep, and I certainly had no place calling him a hypocrite, but there were a lot of blurry lines when it came to this process of ascertaining what was sustainable and what wasn't, especially in terms of diet. A lot of vegans will criticize their omnivorous companions for choosing food that requires the abuse of animals and the land, and yet, eating only plants might also mean that a few field mice get killed by a tractor as a farmer reaps his millet harvest for their breakfast cereal. It also means relying on vitamin supplements like the B_{12} in nutritional yeast, which doesn't occur naturally and therefore must be manufactured synthetically, not to mention importing spices from across the world via plane, train, boat, or truck. Furthermore, if you're not on a raw diet, there's the energy required to heat up all the veggie stir-fries and tofu curries.

This isn't to say, of course, that vegans leave as much of a carbon footprint as meat-eaters—far from it. But it seems impossible to eat anything without leaving some impact on the earth, so perhaps we should at least own up to this, regardless of whether we choose to eat beef lasagna or raw kale for dinner.

AUGUST 13, DAY 166

Keep a brick or plastic bottle in the toilet tank

"Can we just talk about composting toilets for a second?" I asked Andrew, an elongated twenty-year-old version of Paul Rudd. I developed a bit of a crush on him during the bike trip—mostly because he was tall and Paul Ruddy, but also because he had brought his own vintage Bianchi road bike with him instead of renting one of the company hybrids. He had an English lit degree, too, so was guaranteed to have more than just *Silent Spring* and *The Omnivore's Dilemma* sitting on his bookshelf. Plus he played weird string instruments and had a good voice, and—less importantly, but secretly most importantly—he wasn't vegan. Not even close.

The subject of composting toilets, I quickly realized, may not have been the best one to pursue if I was hoping to make any romantic progress with Andrew. Unfortunately, it was too late. I'd already blurted it out.

The best tack, I decided, would be to look cute and somehow segue into another topic. Unfortunately, there's this problem I have whenever I'm in awkward situations: I say something stupid, realize I'm saying something stupid, tell myself to shut up, and then proceed to say something even more stupid.

So while the voice in my head was saying, "Stop talking about composting toilets, stop talking about composting toilets, stop talking about composting toilets," the voice coming out of my mouth was saying, "I can't get over how they don't smell!"

"Yeah, it's pretty cool," said Andrew, gracefully. And to be fair, of all the things on the bike trip so far, the only thing really blowing my mind was the composting toilets. When Mark had asked all those questions on the first day, there was just one to which I couldn't reply in the affirmative, and it was, "Who here has used a composting toilet?" At the time, I thought such a contraption would involve worms, a hole in the ground, and the ability to maintain a squatting position for an extended period of time, so when he said we'd be using them at Harry's farm and at Dharmalaya, a yoga center that would make for our final campsite in Eugene, I prepared for some hygiene anxiety and made a mental note to avoid eating fiber during this time.

But in actuality, these toilets—which can be as simple as a

bucket filled with sawdust or as complex as a self-aerating, temperature-controlled unit—are incredibly sanitary, odorless, and don't require any water. After filling up to a certain point, the waste is then turned into fertilizer, either in a separate, heated compartment or an outdoor compost unit. This "humanure" can safely be used on the garden after a year or so, when it's become sterilized. Using these things, I was astounded at how those portable toilet companies could still be in business. Why would construction crews or concertgoers want to bother with all that rancid blue liquid, toxic air fresheners, pump mechanisms, and so on, when all they really need is a bucket and some sawdust?

At home, before flying out west, I had tossed a water bottle in the back of my toilet tank in advance for today's change—doing this meant that many milliliters of water got displaced each time the tank filled up and less got flushed each time. I couldn't notice a difference when I tried it out, so it was an easy and successful enough change, but now I was convinced that water was for suckers, and if there were any way to install a composting toilet in my apartment, I was going to do it as soon as possible.

"It doesn't even look gross," I heard myself saying to Andrew. God, I was still going on about all this? Time for a different approach. I switched into journalist mode and started asking open-ended questions, ones that began with "How do you feel about . . ."—and ended in anything other than ". . . composting toilets?"—hoping this might open the door for more tangents, which in turn might lead to less crappy conversation subjects, such as the importance of organics and fair-trade, or even better, how pretty I look today despite my Spandex bike shorts and bandana, or, the ultimate best-case scenario, how chilly the nights are getting here and how thin my sleeping bag is and how warm and pleasant it would be if we just cuddled a bit in his tent tonight.

Luckily, while I may be a spastic conversationalist with no ability to control my potty-mouthed ramblings, I am a decent journalist, and my plan worked. That evening, I slept in Andrew's tent—and no, I wasn't naked. It was far too cold, and besides, there were two other people lying right in between us. It ended up being an uneventful night, but in the morning, when everyone was off getting breakfast, I shuffled and rolled over to him. He asked if I was still cold and I replied that I was, chattering my teeth for emphasis. So he twisted one arm out of his sleeping-bag cocoon, wrapped it around me, wrapped it tighter, and soon enough we were holding hands, nuzzling, and finally kissing. It was a funny thing, having a first kiss before either of us had really woken up, brushed our teeth, or rubbed the crust from our eyes. But it was sweet and tender and quiet, this little tryst—nothing intense, nothing hurried, nothing below the belt. It reminded me of the fact that my relationships with real people are just as important as the ideological relationship I have with Gaia, whether it's a shared belief in the miracle of composting toilets or a shared kiss in a warm, rustling tent.

AUGUST 14, DAY 167

Buy only ethically made, sustainable footwear

I will not buy Birkenstocks. I will not buy Birkenstocks. I will not buy Birkenstocks. Although, I will admit, they somehow look less hideous than they did six months ago.

AUGUST 17, DAY 170

Skip gown at doctor's office

I'm starting to wonder exactly how many of my changes will involve getting naked and how I'm going to feel about this come winter. Maybe I should interview a nudist and conduct some serious field research.

Then again, my mother's opinion would probably suffice.

It was her birthday today, which meant Emma and I cooked her dinner at home—or rather, I cooked while Emma offered her critique from the sidelines—and followed it up with some gift-un-wrapping and cake-slicing. I figured it was as good a time as any to ask about gown protocol in the office.

"You know, I've had patients choose to go naked before," she said, "and it made me feel pretty uncomfortable. Usually it's just the occasional German woman who's used to doing that in Europe and is totally fine being seen naked, but it's still a professional environment. We're taught in school that we should always cover up the parts of the body we're not examining, and I'm sorry, but if I walk into the room and you're just sitting there completely naked, there's kind of a sexual tone to it."

She asked me if I would be as willing to forgo the gown if I had a male doctor, and I had to admit, I probably wouldn't.

"Although," she said, "I should add that if you're a male doctor doing a pelvic exam on a female patient, you're expected to have a nurse there in the room, too."

Now I was feeling some remorse. I'd had a doctor's appointment that morning, and when she said I needed to take everything off in order to get properly weighed, I just stripped down to my underwear right there and then. It didn't seem awkward or anything and we went on making casual conversation about what films we'd seen recently. But after hearing my mother's point of view, I gave it more thought and decided that the next time I had a doctor's appointment, I was somehow going to cram a bathrobe into my purse that morning, to use for both the environment's and modesty's sake.

AUGUST 23, DAY 176

Get massages by people, not chairs or electronic gadgets

I just stumbled upon this website called True Green Confessions, where you can type in an anonymous eco-sin or controversial belief or what have you, and I'm loving it. It's a reminder to us all that hippies are multidimensional, too, and don't always relish the task of turning compost or sorting the recycling. Another cool feature of this site—which I've been trawling for the past hour or so, now that I'm back in Toronto and the glorious world of the Internet—is that, at the bottom of every confession, there's a little "Me too" button you can click to show your empathy.

For the most part, the entries are pretty lighthearted.

Some examples:

"I think the Prius looks like a doorstop."

"I hate it when I accidentally buy vegan desserts."

"Sometimes I quietly slip a newspaper or catalogue into the garbage can instead of the recycling bin, and damnit it feels good."

"I make my kids save energy by telling them to turn off all the lights, but I still sleep with a night light."

"Cloth diapers suck."

"I throw batteries in the trash."

"I use one plate to microwave food on and then transfer it to a second plate to eat from. I just hate the puddle it leaves on the first plate! Then I throw it all in the dishwasher."

"I flush the toilet twice for every poop."

"I buy recycled paper for the guest bathroom but then I get the non-recycled cushy stuff for mine."

However, there are also some confessions that are more depressing, even disturbing.

"I think the green movement is an Orwellian nightmare."

"I'm so tired of being as green as possible and everyone else not giving a crap."

"Why should I care about saving the world when I can barely save myself?"

But my favorite of all the ones I read had to be this one.

"I feel very smug when I walk out of Shop Rite and every single bag I have in my cart is an earth-friendly canvas bag."

Aha! A hippie who admits to feeling smug every now and then, thus making the tag line on my blog all the more appropriate. Wait a second, now I was feeling smug about trying not to be smug. Shoot.

I decided it would only be right to leave a confession of my own.

But what to write? There were so many things I felt guilty about, so many eco-friendly chores I was already sick of doing, I didn't even know where to begin.

Then, it came to me.

"Dear True Green Confessions," I wrote. "Forgive me for I have sinned. The other week, I could be found in the sacred backyard space of a yoga sanctuary in Eugene, Oregon, making out with not one, but two hippies—separately, but both within the span of 12 hours—with little to no regard for my fellow tent mates, who were most virtuously abstinent and most likely vegan."

Yes, it's true.

I never do this kind of thing, I really don't. But Andrew, who had been on the bike tour for two full weeks, left a day early to go back home, and that night Mark arrived to stay over because he had to give another orientation at Dharmalaya the next morning. Back when he was leading our group in a question-and-answer game about active and passive solar, I had remarked upon his nice tan, blue eyes, and unfortunate Polynesian wardrobe, but when I realized he wasn't coming on the trip, I suppressed any romantic sentiments that might have cropped up. Seeing him at the end of the week, though, was different. He walked up to the group—we had congregated in a circle outside, in the dark, with our sleeping bags and flashlights—still wearing Tevas but decked out in sportier camping attire up top, including an LED headlamp, which not only compressed his poofy hair but left his hands free to carry a large jug filled with some sort of locally made vegan dressing that had a label bearing the words *Yum Sauce.*

"Yum Sauce?" I inquired.

"Yeah," he said, shining his headlamp in my eyes. "It's really yummy."

"You don't say."

He twisted the light upward and we kept talking—about the sauce, about the cold, and about how it was somewhat pathetic that despite my Canadian upbringing I'd managed to pack only an "equatorial" sleeping bag that could easily double as a summer shawl. Looking at Mark standing there with his headlamp and Yum Sauce, all I could think was: this dude is superdorky, alarmingly skinny, and victually prepared. Exactly my type.

Maybe it was the lactic acid–induced delirium at the time of finishing a 130-mile bike ride, maybe it was a desperation in my gut as a result of meat, dairy, and caffeine withdrawal, or maybe it was just a general high from a week's worth of organic energy bars, sustainable dialogue, and composting toilets—whatever it was, it took hold, and less than five minutes into this inane Yum talk, Mark and I could be found climbing into the closest tent together. Shameful? Yes. Embarrassing? Yes. Fun? Also yes.

The next morning, everyone was heading back to Portland and going their separate ways. Most people took the train but Mark had a car and, well, sleeping with the orientation leader had its benefits: I got a free ride. At first I felt guilty about indulging in this mode of transportation, but he was going to be driving anyway, so really I was just helping him carpool.

On the way back, I stared out the window, reflecting on the sheer absurdity of this situation, how I'd set out on this journey inwardly condemning all earnest vegans but at least looking forward to some cycling and recycling, then a week later found myself hooking up with one who personally chauffeured me back to where I began. Frankly, I was exhausted, and the aches and pains I'd been suppressing with a continuous stream of Advil were now making themselves known. I let out a sigh and cranked my head to one side. Just then, Mark reached over and placed his thumb and middle finger on the upper nape of my neck—with a single squeeze I was suddenly flooded with ease, my blood rushing through the broken dam of stress and anxiety.

"Holy shit," I said, breathless. "I completely forgot you were a massage therapist."

Mark smiled, fully aware—blasé, even—about his superpowers.

"Oh my god, wow," I said. "Seriously, wow. How did I not realize you could do this? And can you do this indefinitely? And will you marry me, please? Let me know if I'm moving too fast."

Part of me actually meant it, too. At this moment, it was as clear as the blue sky above that the world would be a far better place if we all just stopped what we were doing and gave one another a massage. A real massage. No coin-operated chairs or battery-powered gadgets or Jacuzzi jets—a straight-up, human-powered massage. This, more than any solar-paneled roof or vegan diet, was definitely the best way to save the planet.

AUGUST 24, DAY 177

Choose an electronic rather than paper ticket

"I got you a paperless ticket to the Bicycle Film Festival," said Meghan over the phone this morning.

"A what? To which? And when?" I asked.

She had just found out about the Bicycle Film Festival, which is what it sounds like: a film festival with a cycling theme. And when she saw that the tickets were being sold through a website called BrownPaperTickets.com—which, despite its misleading name, gives you an electronic receipt instead of a paper one—she decided it would be a good green change for me. I agreed.

I also figured, being a film critic and a cyclist, this was an event I should check out anyway. Meghan isn't such a cinephile, but she does love riding her bike—a silver one she named Betty, which has two large baskets, both of which are exploding with a mess of colorful plastic foliage, clip-on butterflies, various lawn ornaments, a personalized mini–license plate, and a pink flagpole. Not so surprisingly, it's never been stolen.

We arranged to meet outside the theater, and when I arrived, I was pleasantly surprised to see there was valet bike parking. I handed Quentin over—that's my bike's name—and picked up my claim tag. It wasn't hard to find Meghan in line; it never is, really, because despite her five-foot-one height restriction, she's always dressed much like her bicycle baskets: in a rainbow of psychedelic textiles and accessories. On top of this, at least at the moment, she also had on her paisley-print, neon-green helmet.

"Hey," I said.

"Hey!" she said, taking a swig of water from her stainless steel Sigg bottle.

"Oh no, I forgot mine at home," I said. Suddenly I felt parched, but I was also too embarrassed to ask Meg if she'd share her tap water with me. "I think I'm going to have to break my pledge and get a bottle of Evian or something. You won't tell, right?"

I'd had a series of slip-ups recently—indulging in a late-night falafel that came with paper napkins and a wax-paper sleeve, using my hair dryer more than I should be, eating popcorn with non-organic butter melted on it, to name a few—and was starting to get paranoid that I'd be caught somehow.

"Don't worry, your secret's safe with me," she said. "But hey, you know what you could do?"

"What?"

"Well, when you're done with the water bottle, cut the top off and use it as a funnel when you're pouring your bulk bags of spices into your Mason jars. That's what I do. It makes less mess that way."

Oh my god, I thought. She's even keener about this whole green challenge than I am.

"You are such a nerd," I said.

"You love it."

I left her for a few minutes to find a convenience store and by the time I came back, the line had started to move. Meghan was batting her eyelashes at the valet attendant but I wasn't in the mood for any flirty games—I just wanted to get inside the dark theater so no one would see me sneaking sips from my disposable, estrogen-leaching, imported plastic evil.

AUGUST 26, DAY 179

Eat only organic tubers and root vegetables

At last, the cottage. Time to relax and do nothing (except update my blog every day and conjure up more green changes, of course—a quotidian responsibility that was beginning to grate slowly but piercingly on my nerves). Mom and Dad had rented a cozy, secluded cabin on Hope Island, up north in Muskoka, so that's where I was headed for five days. My diet would consist of nothing but Canadian beer and fair-trade s'mores (well, the chocolate at least), my exercise would be limited to tanning on the dock, and my social life would be confined to the pages of Salman Rushdie's *Midnight's Children*, which easily beat *Anna Karenina* once again in the what-book-do-I-take-on-vacation showdown.

After a couple of mix tapes and a few wrong turns in my mom's car, I finally made it to the dock, where my sister was waiting with the boat to take me over.

"Heyyy," she said—or practically sang—when she saw me. She seemed to be swaying a little more than usual as she walked and didn't seem at all concerned that stray wisps of hair were dancing out from the top of her French braid. This is unusual for Emma, but then again, she had been up here for a while already and Ontario cottage country can indeed have this drastic an effect; it's like Valium for the soul.

We soon pulled up to the island's shore and I climbed out onto the dock, promptly collapsing on the warm slats of wood, sprawling out in pure contentment. I'd been in the Middle East, Spain, and Oregon, spending hours upon hours outside, but this was the first time I'd truly felt the sun. I closed my eyes, inhaled the breeze, and heard my mother chirp, "Hello."

Regaining consciousness, I made my way up the cottage steps, dropped my bag in the front room, and went to inspect the contents of the fridge. My parents had been good up to this point about feeding me organic, free-range, and local meat whenever I'd gone to their house for dinner, but with the nearest grocery store half an hour's boat ride away, they'd slacked off. The milk was commercial, as were the sausages, which were packaged in Styrofoam and plastic wrap, and I had no idea where any of the fruit or vegetables came from. This was a problem. I had just today sworn off all nonorganic tubers and squash, thanks to a half-hour lecture from Harry at Sunbow about how these deep-rooted vegetables can suck up leftover traces of DDT from the soil far more than any grain, fruit tree, or leafy vegetable.

"I got you some organic shortbread cookies," said Mom, coming up behind me and opening the pantry.

Cookies. Great.

I mean, yes, I was excited for s'mores and everything, but could I really survive on a diet of cookies for five days? My system was just getting used to stuff like quinoa and kale—what if I went into a diabetic coma? I decided to compromise by eating the fruit, vegetables, and bread but abstaining from the dairy, meat, and eggs.

This was fine, for the most part, and my first twenty-four hours at the cottage were total bliss. Soon enough, however, my restless personality and need to do stuff kicked in; bliss morphed into impatience and I wanted company. Most of my friends had to work or were on vacations of their own at this point, but Jacob had just returned to Toronto from Ramallah for his regular summer visit, so I called him up one morning and used every rhetorical argument I could to persuade him that a trip to Muskoka was absolutely necessary: 1) free booze; 2) time-out from a hectic social calendar; 3) rugged, inspiring Canadian wilderness; 4) stimulating conversation with one of his best friends; and 5) um . . . organic cookies? In hindsight, I probably could have convinced him with argument 1 alone, but anyway, he agreed and I picked him up from the dock the following afternoon.

We spent the next couple of days doing pretty much what I'd been doing up until then—that is, nothing—but somehow, his presence helped. A lot. He made the conversation a little more engaging, the movie-watching a little cozier, and the meals a little more . . . well, dairy-free, I suppose (the lactose-intolerance thing).

On the second-to-last day at the cottage, however, my stress began to kick in again as I remembered I hadn't yet thought of a blog change beyond tomorrow, let alone anything for the coming week. So when the afternoon sun made our skin ache for shade, Jake and I went inside and sat down at the kitchen table with my parents and sister, and we all began brainstorming.

"No leaf blowers!" said my mother. But I didn't have a lawn.

"Stop breathing!" said Emma, most unhelpfully.

"Get a crank-up radio—I saw one at Roots the other day," said Mom. Not bad, but that would involve buying something new, which wasn't great.

"Stop eating!" said Emma. I let out a forceful sigh.

"Renting DVDs instead of buying them?" said Jacob. That was doable, although I didn't often buy DVDs to begin with.

"Could you commit suicide and green that somehow?" asked Emma, again unhelpfully and now somewhat frighteningly.

"Have you given up paper napkins yet?" asked Jacob.

No! I hadn't! Perfect! I thanked him with a warm, suntan-lotion-sticky hug (that would be natural, preservative-free suntan lotion, of course) and wrote down my next change.

AUGUST 27, DAY 180

Online dating with GreenSingles.com

"I think Jacob has a crush on you," said my mother, after he'd left the cottage to go back to Toronto, and eventually all the way back to the West Bank.

"What? No, no, no. No way," I said. Really, there was no way. If anything were to ever have transpired between us, it would have happened at least once in the past decade and a half that we'd known each other. Besides, he lived halfway across the world, and chewed with his mouth open sometimes. I mentioned these two clearly irredeemable facts to my mother.

"He does like you, though, I can tell," she said. "Why don't you two try dating?"

Had she lost her mind?

"Uh, no, Mom," I replied curtly. "I can't date Jacob."

"Why not?" asked my sister, who swiftly strolled into the room, cracked open a beer, and chimed into the conversation in less than three seconds.

"He's smart, he's skinny—what's the problem?"

It's true, he was technically my type.

"You don't understand, though," I said. "I've known him for fifteen years, we've been nothing more than friends all this time, it would be weird if we got together—and besides, you're wrong, he doesn't have a crush on me. Please."

"He does," said my mom. "And knowing someone for a long time isn't a reason not to date them. Why not just consider it?"

"Oh my god," I said. "Because . . . because it's Jacob." And we left it at that.

But I decided, at this point, that it was definitely time to get more proactive about my lack of a boyfriend (or really, just lack of any dates in general) and perhaps dabble in the online dating scene. That said, there was no way I was about to pay a cent for anything, nor did I particularly want to trawl through page after page of skeezeballs from the 'burbs looking for nothing but "a good time" at the nearest nightclub. Then I found Green Singles.com, and I realized I could kill two birds with one stone: find a boyfriend and call it a green change. At least by using a service like this, I'd know the guys would have already passed the first test—that is, they'd understand things like my love for farmers' markets and my hatred of Styrofoam. I was hoping there might even be some normal guys who would also understand my love for Advil and my hatred of hemp necklaces.

The website, which was for the "environmental, vegetarian and animal rights community," had been around since 1985, so that was a good sign. But it also had an inspirational quote of the day, which was "Sustain me with raisins, refresh me with apples; for I am faint with love." Song of Solomon 2:5. Ick.

I created a profile anyway, which was probably the most painful and embarrassing thing I've ever done in my life. Considering I'm a writer by trade, it came as a surprise how challenging it was to sound normal. Reading it over, I realized I came across as utterly neurotic and high-maintenance—why would anyone respond to this ad? I certainly wouldn't.

But it was too late in the evening and too far into a bottle of Ontario Baco Noir to bother finessing it, so I moved on and did a search, selecting men looking for women, in Toronto, age twenty-six to thirty-nine. The number of matching profiles that were eventually returned: seven.

The first said he enjoyed reading "Doulgas Coupeland." Nope.

The second was someone with the headline "Walked out of the wilderness all squint-eyed and confused." In his "About Me" section, he wrote, "I have lots of friends but am looking for love. I have been a good vegetarian/bad vegan since high school, have strong political views in the anarchist, environmental and anti-oppression viens [*sic*]."

Also, nope.

Then there was a guy whose profile picture was of him in a yoga pose. His caption said he was looking for yoginis, and under religion he had chosen—from the drop-down menu, mind you—"On a spiritual path." He wrote that he'd like to meet a woman who "is aware of her connection with the divine. Takes care of her body. Possibly has been to India or wants to go. Knows what the word kirtan means. Knows exactly what she wants, and has no doubt that who I am fits into that want."

Ugh, definitely nope.

The last profile I clicked on had the headline "Nature boy seeks earthy wild woman who loves art, outdoors and gliding."

There was something funny about how this guy included general loving criteria along with an ultraspecific gliding reference. If he met a woman who was a beautiful, eccentric, woodland painter but didn't love gliding, would it never work?

I checked to see what he'd written under "About Me." It said, exactly: "I'm all about CONNECT-I-on. Nature is my life. I love to stop, watch, touch. And any activity outside keeps me leaping: organic gardening, kayaking, cycling, hiking, tai chi, meditation, swimming (I am a fish) in abandoned quarries and oceans ... Compassion (feel with) fills my life."

Then under the "About You" heading, it said: "I want someone in touch with her feelings, especially her fears. She should be working on dealing w/ them and FACING them (usually linked w/ childhood/family). I also want someone who likes to connect, collaborate and work two-gether."

You'd think I would have stopped reading by now, but no. I just had to know what this punctuation-obsessed guy listed under "Interests and Hobbies." And it was, indeed, a real gem: "Yoga, kundalini, deconstructing, playing with my chickens, fermentation, using my composting toilet :)"

Whoa, BIG NOPE.

I mean, I love the composting toilets, but NO.

Forty-eight hours later, I took my profile down, poured myself another glass of plonk, sat next to Sophie on the couch, and closed my eyes.

This whole search for eco-friendly companionship was beginning to frustrate me, to the point where I was almost ready to give up. But part of me believed that if I were trapped in the Heartbreak Hotel and my true love was waiting for me on the other side of the door, perhaps the answer wasn't to keep pushing and pushing at it, but instead, try pulling. After all, it's often the simplest solutions that prove to be the most effective—in fact, maybe I've already found The One; maybe he's in the room next door and I just haven't realized it. My head was ringing with all these thoughts, when suddenly something else started to ring.

My phone.

"Hello?" I said.

"Hey," said a voice. A languid, male voice.

It was Mark.

When we'd hugged farewell at the airport in Portland a couple of weeks ago, I thought that was the last I'd see of him—my summer fling with a hippie, nothing more. But then he sent me a few lengthy e-mails about my green challenge and mentioned how he'd liked hanging out with me, as well as a couple of text messages just to say hi while I was up at the cottage. I hadn't bothered replying to these because, for one, the phone signal was lousy on the island, but for two, Jacob and I had been so preoccupied with our cannonball competitions, midnight conversations on the dock, marshmallow experimentation, and so on. Though Mark was a nice guy, the truth was, when I was reminiscing with Jake about some catastrophic semiformal we had in the eleventh grade, or delving into issues like cross-cultural empathy and the possibility of true happiness, or just snuggling up with him on the couch, watching *The Shining* for the eighth time and feeling utterly at ease in the knowledge that this was a friendship with an infinite life span, the last thing I wanted to do was type a strategically worded text message to a guy I'd never see again.

And yet, here I was, talking to Mark, thinking about how he was a lot more normal than any candidate on GreenSingles.com. Near the end of the conversation, he said something about a trip he was planning to the East Coast to visit his parents, and the possibility of flying up to Toronto on his way back. No pressure, he said, but would I be interested in seeing him for a few days, maybe even a week? It took me a few seconds to reply—I couldn't even remember the last time I had a cross-country booty-call request, let alone one from a thirty-nine-year-old vegan in Oregon.

"You know," I said, finally, "I think I would."

AUGUST 29, DAY 182

Make jam and preserves

Making jam has always sounded like an incredibly mundane activity to me, along the lines of buying new watch batteries and descumming the shower curtain. I also think of jam as something you give to someone when you have no idea who they are and you don't really care; like, "Hey, here's a jar of toothaches . . . It might make your toast less dry, at best."

But today, in the name of preserving food and preparing ahead for the scurvy I may or may not be enduring come winter—when I'll probably be limited to food grown within Ontario and eating cabbage for breakfast—I made jam.

The recipe and instructions, which I got from Crunchy Chicken, a fellow green blogger living in Seattle, were surprisingly easy to follow, most likely because I didn't so much follow them as make them up. I wasn't sure what blanching was, nor did I have any idea where pectin came from or whether it was environmentally sound. But somehow I managed to get all of the ingredients together, throw them in a pot, and bring them up to a gentle boil, the maximum my green rules would allow. I stirred the mixture around until it started bubbling and turning gloppy, poured it in some Mason jars that I had "sterilized" (that is, dipped momentarily in hot water), and left the whole batch to sit overnight.

But after just a few hours, I couldn't resist and very tentatively opened one of the jars to see what it looked like, half expecting to be confronted with some foul stench and a runny mess, but lo and behold, there was something that resembled jam. My jam. My local peach, organic plum, and unfortunately imported vanilla-bean jam. I scraped some out onto a cracker and bit into it, chewing slowly, preparing my gag reflex for whatever flavors might trigger it, but it actually tasted good. I couldn't believe it. I made jam. Good jam. Now I'd just have to find myself an unsuspecting mother-in-law to pass the other six jars on to.

Then again, maybe Mark could help me with that. After talking on the phone with him, we agreed it would be better if he came to visit sooner rather than later because the film festival was starting in a few weeks, at which point I'd be running around the city for fifteen hours a day and too stressed out to play host. So he booked a ticket to see me first, before going down to North Carolina to see his family, and arrived last night.

I picked him up from the airport in a Zipcar, opting to pay the few extra dollars for some specialty wheels, i.e., a red Mini Cooper, which I thought would look pretty impressive. He was easy to spot, with his enormous backpack, waist pack, and laptop case, and when he saw me, he gave me a kiss, which would have been nice had he not opened his mouth so wide and smelled so much like a plane cabin. Still, I was happy to be picking up a cute boy from the airport and taking him back to my apartment. As we began walking over to the parking lot, I prepared him for the impending coolness of our ride home.

"I rented a Mini," I said, grinning through my eyes. "It's supercool, so you better not try to be nerdy in it or anything."

"What's a Mini?" he asked.

"What do you mean, 'What's a Mini'? You know, a Mini—like, the car."

"A Mini's a car?"

What the hell?, I thought. Who doesn't know what a Mini is? I mean, I don't know cars, but this is an icon—you only need to have seen a James Bond film or two, or *The Italian Job,* or even just *Austin Powers* to know what a Mini is.

Weird.

Anyway, we got out of the parking lot and onto the highway.

As we drove, he rested his hand on my knee, a gesture that was much more appreciated than the sloppy kiss at the airport, and I felt good. I'd brought him a slice of vegan carrot cake, too, so when we got home, he showered, unpacked, and took a few bites of that while I had a few sips of wine, and then we promptly got down to business (the business of making out, that is).

It was just what I needed—from the talking to the eating to the rolling about on the couch, it felt as though I were finally injecting some life into my living room. Mark was good, too: caring, thoughtful, honest, and sweet.

Not as sweet as my jam, mind you, but that's hard to compete with.

september

1	Use revolving doors
2	Groom my cat regularly to prevent hairballs and shedding
3	Order pints rather than bottles of beer; get kegs for parties
4	Use a natural lubricant instead of K-Y
5	Wear clothes twice before washing
6	Write smaller and use both sides of each page
7	No toilet paper for number one
8	Exterminate bugs with eco-friendly solutions
9	Use a natural myrrh-based mouthwash
10	Limit computer use after dinner
11	Rent DVDs instead of buying them
12	Decrease margins on Word documents
13	No more birth control pill
14	Sign up for a biweekly organic CSA
15	Buy ceramics only from potters who recycle their clay
16	Reuse envelopes
17	Use towels at least five times before washing
18	Eat only ethically raised fish
19	Cut off the end of my toothpaste tube to use every last ounce
20	Leave the DO NOT DISTURB sign on the hotel door for duration of my stay
21	Use incense rather than artificial air fresheners
22	Use chalkboards rather than flip-pads for presentations
23	Water my outdoor plants at night
24	Shave my legs in the sink rather than the shower
25	Boil water and sauces gently, not at full heat
26	Use an eco-friendly cutting board
27	Use a natural saline solution instead of Visine eye drops
28	Use cloth menstrual pads instead of disposable ones
29	Eat ice cream in a cone rather than a plastic cup
30	Stay on the path while hiking

SEPTEMBER 1, DAY 185

Use revolving doors

Apparently, according to some smarty-pants at MIT, regular doors exchange roughly eight times as much air as their revolving counterparts. The study's authors explain it like this: "Our estimates show that if everyone used the revolving doors at [the university building] E25 alone, MIT would save almost $7,500 in natural gas amounting to nearly 15 tons of CO_2. And that's just from two of the 29 revolving doors on campus."

I'll also bet that even more energy is saved if two people try to squeeze into a single quadrant of a revolving door—after all, puerile hijinks can be green, too.

SEPTEMBER 4, DAY 188

Use a natural lubricant instead of K-Y

Despite my various control-freak habits and need for boundaries and personal space, I'm actually quite happy to live with another person, provided that at least 80 percent of our idiosyncrasies are compatible and basic levels of hygiene are met. In this case, there were no problems—Mark is clean, he sleeps and wakes at reasonable hours, he's up for doing anything, but also nothing, and of course the daily massages helped release any tension that might have accumulated. Furthermore, abiding by my green rules wasn't a problem, as he was already living by most of them anyway, and when I went to work each day, I left him with single-spaced, small-margined notes suggesting things to see and do. One day, when I was working from home, I was concerned he'd get on my nerves. But instead he just sat quietly on the couch next to me with his laptop, doing his own work. After a few hours, he got up to make a snack—rice cakes with hummus and cucumber—and brought it over for me on a plate. He then cleaned up and sat back down again, gently resting his hand on my foot.

It was domesticity at its best. This was a guy who cared immensely about the earth but also about people, who brought me rice cakes with hummus, gave me massages, and was capable of being tender without being distracting. What more could I want?

And yet, and yet . . . could I really be with someone who refused to drink coffee? Who didn't know what a Mini was? Who thought monogamy was so yesterday?

I decided to do what any perfectionist-cum-procrastinator-who-overthinks-things does, and make a list.

Turning my screen away from him, I created a text file on my desktop titled, cryptically, "Pros/Cons."

After about ten minutes, it read:

Things I like about Mark:

- Excellent massages
- Likes riding bicycles
- Environmentally conscious
- Brings me rice cakes with hummus
- Can fix most computer problems
- Has a great tan and is skinny
- Uses coconut oil as a lubricant
- Good in bed
- Seriously, seriously good massages

Things I question about Mark:

- Pauses a lot before speaking so I'm not sure whether to keep talking and fill in the blank air or wait 30 seconds in awkward silence for his response
- Doesn't read fiction
- Wears Tevas unironically
- Hasn't heard of Joy Division, let alone formed any sort of opinion on whether he likes them, nor has he heard of *The Office*—plus, when I finally made him watch it one night he said he didn't get it, and on top of that pretty much hates all pop culture regardless of the fact that pop culture is my lifeblood
- Breath sometimes smells bad
- Uses coconut oil as a lubricant
- Doesn't like cats; definitely doesn't like Sophie
- Says things like "dang"

I still didn't know what to do.

SEPTEMBER 5, DAY 189

Wear clothes twice before washing

Each morning, after brushing my teeth and having yet another dark, lukewarm shower, I walk over to my closet and decide what to wear. Normally, this involves scanning the racks of pants, opening and closing a few drawers, trying to imagine a combination of clothes that might work together. But I was reminded, with Mark here, of the way most men decide what to wear each day: picking things off the floor and smelling each one to see which is cleanest.

I'm far too neat to leave my shirts anywhere but in the closet, but I am starting to realize there's nothing wrong with wearing something twice, as long as it doesn't stink. Recently, in fact, I wore a sweatshirt three days in a row because I was up north on a camping trip—or rather, Mark and I were up north on a camping trip.

It was my way of figuring out how I really felt about him. The best way to do so, I concluded, was to test him in two crucial ways: one, go on a mini-vacation that involves a bit of time on the road, in a confined space, and two, introduce him to Meghan, Ian, and my family.

We decided, then, to go camping in Algonquin for a long weekend. It would require driving a few hours north, canoeing across a few lakes, portaging, and paddling once more over to an island. Mark stocked up beforehand on food—organic yams, quinoa, dates and almond butter, vegan energy bars, split-pea soup mix, and so on—and brought along his water purifier. I borrowed my dad's car, a Jaguar SLX, which wasn't exactly a hybrid, but the fact that I'd offsetted the drive ahead of time eased my guilt somewhat. It didn't, however, ease my embarrassment when it came time to strap the rental canoe up top (the employees at Algonquin Outfitters insisted on taking a photo for their records). We got lost a few times, but because I get lost almost every time I get in a car, I'm now at least pretty good at knowing when I'm lost, which means we never went very far in the wrong direction. Also, Mark is a persistently calm human being so there wasn't any squabbling, and by the time we finally arrived, we were both in good spirits.

As evening set, we went for a short hike and a skinny dip, arranged our tent, gathered wood for the fire, and prepared a dinner of accidentally burnt yams. Mark's silent streak was a welcome change from all the noise I was used to in the city, and a sense of peace began to descend upon our little island. Sitting there on a damp log, eating a burnt yam, watching the fire kiss the air, and having the moment get officially stamped "Canadian" as a loon called out from across the lake, I felt completely at peace—totally and utterly content. There wasn't any flash of enlightenment, nothing metaphysical or even profound about it, but this was without a doubt one of the happiest moments of my life. It dawned on me then, however, that while Mark was an integral part of this, my deeper sentiments were tied more into the environment—not in that misanthropic fashion or in any sort of "Wow, we're so insignificant in the grand scheme of things" way; it was just a sensation of being outwardly still while inwardly moved . . . a complex simplicity, almost like being embraced from a distance.

I came back from that weekend with Mark firmly believing that nothing else mattered in my life as long as there was someone to make me smoothies in the morning, fix my computer in the afternoon, and massage my feet at night.

Later, after posting some photos online of our camping trip, I received an e-mail from Meghan saying, "I want a cute boy who burns yams for me, too!" I smiled, phoned her up, and asked if she wanted to meet him, which of course she did.

There was a Critical Mass ride coming up, a monthly event hosted in cities around North America in which anywhere from ten to five hundred cyclists gather and ride their bikes, very slowly, around all the major streets in the downtown core. It's an organized thing, though not exactly legal. The point is to raise awareness about the importance of bicycle lanes with the obvious subtext being to give the proverbial finger to motorists, and Meghan suggested we all do the ride together. I'm not big into protesting, but I did pledge to get involved in some sort of activism for my green challenge, and felt I should start taking this beyond my THESE COME FROM TREES sticker campaign, which was mostly limited to public bathrooms. Besides, the only rallying cry I'd be expected to make on this ride would be the sound of my bell ringing every now and then. I could deal with that.

Meghan met up with us that afternoon, and within five seconds she was giving me looks of approval accompanied by a few sly winks. She and Mark quickly launched into a discussion about Ayurvedic body types and were soon comparing notes on rice milk versus hemp milk. Eventually, the ride got under way and the three of us began pedaling down Spadina Avenue, taking photos and pointing out cool bikes. At one point, Mark went ahead to get some more pictures, and Meghan gave me her official thumbs-up, saying as long as I was happy, that's all that mattered.

"I think I'm happy," I said.

A few nights later, Ian called. He had tickets to the Architecture in Helsinki show at the Opera House on Wednesday—did I want to come? I said sure and asked if I could bring Mark, seeing as he was here staying at my place and had nothing else to do. Ian, who's always ready to meet new people—and was now mindful about his readiness to do so—was more than happy to oblige. We met up at his place first and walked over to a burger shop that served organic dairy and grass-fed beef for a quick bite. On the way to the restaurant, Ian launched into a discussion about the aesthetics of postmodern architecture as applied to local housing initiatives; Mark said nothing. Ian then took things down to the level of "This building's ugly, this one's cool," but still, nothing. We crossed the bridge over the Don River and arrived at the burger joint, where Ian and I ordered poutine, onion rings, and beer; Mark, seeing only meat, cheese, and trans fat on the menu and also being averse to alcohol, ordered nothing. On our way to the concert, Ian and I began talking about our love-hate relationship with the hipster community. I noticed Mark would hold his breath and squint his eyes whenever we walked past a smoker. When we finally got inside the Opera House, more beers were obtained, after which Mark asked if the floor was sticky or if it was just his shoes. (I was tempted to tell him it was just his shoes because that's what hiking boots do when you bring them to live music venues.) Ian made subsequent efforts to wring some dialogue from him, poking and prodding with questions about Portland, music, and global warming. Mark said something about how what we really need is an apocalyptic environmental catastrophe to wipe out a significant amount of the world's population and then finally people might be willing to stop driving SUVs and actually live responsibly. Now Ian was quiet, squinting his eyes, opening his mouth, and shaking his head, which was clearly bursting with possible retorts but had no idea which one to fire first. He finally said something, but I couldn't hear it above the blaring cacophony of the opening band. He started to ask Mark further questions, but Mark had seemingly said all he wanted to say and was back to his blank-stare-and-pause trick. When it comes to Ian and me, two best friends who not only finish each other's sentences but can say a single word to convey seven different things and thus are able to wrap up our opinions on a number of multifaceted subjects and all possible tangents in less than five minutes, it's hard to tolerate such nonverbal communication.

In a word: it was a flop.

I was reminded later of something my friend Kieran once said about the East Coast versus West Coast dynamic.

"People from the East Coast are laid-back about being uptight," he said, "and people from the West Coast are uptight about being laid-back."

This seemed fairly accurate, although in truth, Mark probably wasn't so much uptight about being laid-back as he was laid-back about being laid-back, which in my fast-paced, hyperaware world unfortunately comes across as socially comatose.

Ian called the next day and, in his mindfully blunt fashion, said he had fun last night and that Mark was nice, but not good enough. A similarly taciturn—or, really, speechless—dinner with my family a couple of nights later confirmed this, and with a heaviness in my gut, I sighed and made up my mind. As much as I valued Meghan's opinion and would never dump a guy just because my parents disapproved, Ian was right. It's not just about knowing what a Mini is or whether Joy Division rocks, it's about being culturally attuned, having an opinion, and being able to articulate it. Discussions about sustainability and veganism are fine, but there's more to life than just minimizing and dissecting our footprints.

After three weeks, Mark went back to Portland. For good.

SEPTEMBER 6, DAY 190

Write smaller and use both sides of each page

I'm the worst note-taker in the history of all journalists. To prove it, here's a sample of my notes taken during a screening of the film *Persepolis:* "Yuck, Orly. Raised five children with tears. Iron Maiden. If God wills it. Color = present."

In fact, those are ALL the notes I took during that film.

And here, what I wrote during an interview with the director of Luminato, an annual arts festival in Toronto:

"Mill attends 1st. How rapidly T.O. est. itself as centre arts culture. Not ab. passive spect. Light on feet open nite dance part. Lumibateau? MND 8 langs, acrobats, etc. @ least 1 comp +/- artist in res each yr. B/w Harbor and Dist."

Not only do my notes not make any sense, they're jotted down in such a frenzy, I end up scrawling all of four words on a single page, which gets to be a pretty significant waste of paper. This would all be remedied if I had taken the time to learn shorthand in journalism school—alas, my twenty-three-year-old brain at the time decided to watch back-to-back episodes of *The Bachelor* instead, so now I just have a very expensive digital tape recorder.

My accuracy issues are easily solved with this technology, providing I remember to press the record button and keep my batteries fresh, and when it comes to stories that don't require interviews I simply make sure to do copious amounts of research. But there's still the environmental problem—the waste of paper.

I decided for today's change, then, to write smaller and use both sides of the page.

What better subject to test this green change out on than actor Jake Gyllenhaal, who happened to be in town for the film festival promoting his film *Rendition,* which I'd been assigned to review for the *Post.* I wrote my blog entry that morning, giving it the title "Jake Gyllenhaal gets two pages, max," and told my readers, mostly in jest of course, that if they had any pressing questions for the man, I'd do my best to get them answered.

Well, let me just say, if you're a blogger and you want more hits, here's a hint: type the words *Jake Gyllenhaal* into your subject line.

Immediately, my count shot up by about two thousand.

One of the early commenters said, "Wow, not sure you know what you've opened yourself up to here."

Was she ever right. When I checked the post a few hours later, I realized that every fan the actor currently had with a basic working knowledge of the Internet had somehow figured out that I'd written about him, and now swarms of these people—mostly women, who referred to themselves as Gyllenhaalics—had descended upon Green as a Thistle en masse, begging me to pleeeease ask him this-or-that question, which they'd been waiting *forever* to get the answer to. The questions ranged from the straightforward and professional ("What does he hope audiences will take away from his latest film?") to the trite and useless ("What would he tell American youth about living their dreams?") to the utterly, utterly random ("What does he think about cilantro?").

I jotted down a couple of them, just in case I ran out of my own—after all, I didn't exactly love *Rendition* but I did love cilantro—and made my way over to the eighth floor of the Intercontinental Hotel, where the interviews were taking place.

I walked in with my bike helmet tucked under one arm and my water bottle in hand, a film of sweat on my forehead that I wiped up into my frizzy, non-blow-dried hair, and an ill-fitting organic cotton T-shirt with some sort of highlighter stain on the bottom. I immediately felt like a dirty fool standing there in front of Jake, who was dressed in full suit and tie, looking even better in person—and taller, and bluer-eyed—than he does on screen. I should've at least broken my no-gum rule and freshened my breath.

So I did what I do whenever I'm nervous: ramble, ramble, ramble. He was having none of it, though, and interjected to ask what kind of bike I had.

Wait, what bike did I have? Suddenly, I couldn't remember. It was white, yes white, but what make was it? Why couldn't I remember? I just rode it here two minutes ago.

"Uh, an old Bianchi?" I sort of asked. What the hell was I saying? I didn't have a Bianchi, I had a Sirrus hybrid—I mean, I liked Bianchis, especially after getting to know Andrew in Oregon, and was looking at a few on Craigslist the other day, but I didn't have one. Why was I lying to Jake Gyllenhaal? And why did he have to have such a flawless complexion? And such impeccable dental work?

This was getting out of hand. I dug around for my tape recorder, found my notepad, then sat down and got to work.

The conversation plodded along uneventfully for the next ten minutes, but I was his last interview of the day and could tell he was exhausted, bored with answering the same questions about the same movie for hours on end. His quotes sounded rehearsed and uninspired, and my list of questions was looking equally unoriginal. I decided to just go for it and give the Gyllenhaalics what they wanted.

"Okay, listen," I said. "I have to explain something a little odd."

As quickly as I could, then—because I had only fifteen minutes in total, including the time it would take the photographer to get a picture—I briefed the actor on my blog, the post I'd written today about taking smaller notes, and how I'd unintentionally attracted a mob of rabid fans who were just dying to know an assortment of things about him, from whether he still has his puggle, Boo, to his opinion on cilantro.

"Oh, I hate cilantro," he said. "It's the only herb I don't like."

All right. I made an official note about his dislike, muttered something stupid about how it was one of the most divisive herbs in today's society, and went on to ask my next question.

But he cut me off.

"Hey, you're not writing that small."

"Pardon me?" I said.

"Your writing," he said again. "It's kinda big. I mean, you said you were trying to take smaller notes but I can read that from where I'm sitting, even upside-down."

I looked down at the page. He was right, it wasn't very small. But wait, was this really that important?

I lost my train of thought.

"Uh, yeah, I guess you're right," I stumbled. "But it's smaller than usual—I mean you should see the font I usually write in, it's just, like, gargantuan—and I'm using both sides of the page at least."

"Is that a rookie mark on your leg?"

"A what?" I asked.

He reached down and pointed to a greasy chain mark from my bike on my right shin, then smudged it with his finger.

"It totally is. Cool."

Oh my god, Jake Gyllenhaal just touched my shin. He just touched my shin and said cool. Is that allowed? Can I touch him? Maybe he'll do it again. How can I get him to do it again? Why is he staring at me like that instead of answering my next question? Oh my god, I haven't asked my next question yet.

"Yeah, a rookie mark. I always get those. Lube can get messy."

Shit. That wasn't a question at all. That was also the most embarrassing thing to pass through my lips all day. Honestly, I may not believe in God, but sometimes I thank the Lord profusely that my editor hasn't caught on to the fact that I'm really just pretending to be a journalist.

"Um, okay, so you used to have a puggle, right? Boo?"

"Time's up," came a voice from the door. The publicist came strolling in and was calling it a wrap. Great. All I had were some dull quotes about the CIA and the fact that Jake Gyllenhaal hates cilantro. What a scoop. Stop the presses.

As his entourage filtered in, I flipped my notebook shut and packed up to leave. I waved goodbye and thanked him, and he waved back, then gave me a little wink.

"Ride safe," he said.

SEPTEMBER 7, DAY 191

No toilet paper for number one

I'd been toying with the idea of not using toilet paper for at least a few weeks, and in all honesty I wasn't hesitating out of squeamishness or modesty but rather out of logistical confusion. It seemed, if one chose not to use the stuff, there were two options: water or cloths (or some combination of the two). Reading some of the other green blogs, I noticed that certain folks, like the ones following Crunchy Chicken's Cloth Wipe Challenge, were comfortable using handkerchiefs or cotton wipes. But a lot of other people liked to use a bidet—whether an attachment that hooks on to the side of the toilet, such as the TushyClean, or a completely separate "bum fountain" (as one person so candidly nicknamed it). This, to me, sounded much cleaner than the cloth route, at least in terms of dealing with bodily function number two. The idea of collecting all that bacteria and letting it sit and fester in some basket, then washing these dirty things with all my other clothes, towels, and bed sheets sounded completely gross and not in the least bit sanitary. Plus, it would mean using hotter water for the laundry cycle, which requires more energy.

I thought the water-splashing technique sounded better but wanted to ask my mother about it just in case, seeing as she not only knew about the dangers of bacteria but also owned a bidet.

"Do you actually use that thing?" I asked her, on the phone one afternoon.

"Well, not very often . . . Kitty probably uses it the most."

Kitty was my parents' cat—a small, gray, and absurdly fluffy creature who still hadn't learned her own name (or was at least boycotting it, which is somewhat understandable). One of her favorite napping locations was in the shallow bowl of the bidet in the upstairs bathroom.

"It's nice, though," Mom added. "I mean, it does get you clean. And you really shouldn't be throwing anything with fecal matter into the regular wash, that stuff is impossible to get out, unless you use a lot of bleach."

I didn't have a bidet, though, and couldn't afford any of the fancy conversion kits. Then I remembered Colin of No Impact Man saying that he didn't use any toilet paper. He also wasn't buying anything new, though, so what was he doing?

I sat down at my computer and began an e-mail to him.

"Dear Colin," it read. "I have a rather personal question for you."

From here, I went on to explain that I was considering giving up toilet paper in exchange for a water-based method but wasn't sure how to go about doing this. In a roundabout way, I asked if he wouldn't mind telling me what he used, and how he used it.

He wrote back promptly but with strict orders that his reply be "off the record," which I learned in journalism school doesn't really mean anything, at least in that if you choose to say something, you've said it. However, I also understand that when it comes to the minutiae of one's scatological routine, it's perhaps best to paraphrase.

In essence, then, No Impact Man said that he used nothing more than a bowl of water and his hand. His wife and young daughter did the same.

Okay, I thought. This was maybe doable. But did you clean the bowl out afterward? What about potential splashing of dirty water? And if you used your hands this would mean they'd be in contact with poo and, immediately after, in contact with the tap handles, and how long could E. coli survive on such surfaces anyway?

I asked these questions and more to Colin, but he gave me diddly-squat in return, other than the blunt reply that many people do this in many countries around the world, so I should just suck it up and do it, and not get all squeamish about it.

Yeah, right, I thought. Squeamishness aside, how many people around the world also had dysentery because they wiped themselves with their hands? Even in a first-world country with access to soap and hot water, I'm still not confident that my attempts at sterilization are even remotely close to being acceptable—perhaps if I invested in an autoclave and buckets of Purell, I'd feel more comfortable, but then that's hardly eco-friendly. Besides, how much damage could a few extra squares of recycled toilet paper really do to the earth?

In the end, I decided to go halfway and continue using toilet paper for number two but not number one. Urine was pretty clean, and as long as I remained relatively hydrated it would be mostly just water coming out anyway. But this bowl-of-water thing Colin did still wasn't appealing to me, so I went upstairs and rummaged around in my closet to see if there might be something else I could use, something that could maybe shoot out a steady, easily directed stream of water. In the back of my mind, I knew my parents probably had a few water guns lying around the sheds in the backyard, but something about pointing a gun—toy or otherwise—at such a sensitive region disturbed me. Then I saw it: a water bottle I'd gotten for free at the Green Living Show, which had a vacuum-pressured squirting mechanism. That depressing convention was coming in handy, after all.

I could use that, along with a washcloth or handkerchief, which would probably only need to be changed after a couple of days, providing I wiped strategically.

Granted, this wasn't the sort of change I could take with me outside the house—another water bottle and multiple cloths really wouldn't fit in my purse, and the last thing I want is to end up blowing my nose in the wrong piece of damp fabric—but it would at least suffice for at-home use. So I went over to the bathroom tap and filled up the bottle, tightly securing the lid. Then I placed it down on the floor beside the toilet, with one of my older, faded washcloths on top.

Makeshift bidet: $0.

Not bad, I thought, as I walked downstairs and poured myself a big stein full of water.

SEPTEMBER 13, DAY 197

No more birth control pill

I'm sure my mother will be happy to hear about my decision to stop using the pill as it brings me one step closer to getting knocked up. Actually, it was she who first put me on it, back when I was sixteen and dating a guy three years older—dating him, if you ask my mom, purely out of rebellion; except that it really wasn't rebellious at all because he happened to also be three times nerdier than I was. But now that Mark was out of the picture, I was officially single and figured that taking a drug I didn't technically need, other than coffee of course, which I do, in fact, need, was somewhat silly. As well, I had now been steadily altering my hormones for about twelve years and thought it might be time to remind my uterus and fallopian tubes they still have some work to do.

On top of all this, however, is the green factor: I've been reading more and more studies about hermaphrodite frogs and transgendered fish that are a direct result of the estrogen levels in rivers and streams. This is due, scientists argue, to the fact that when women use the birth control pill, some of the estrogen gets absorbed by the body but the majority is excreted in a nonactive form in urine; then, when it gets flushed down the toilet, it hangs around in sewage treatment plants, which aren't properly equipped to filter out pharmaceutical matter. It becomes more and more concentrated, getting mixed with other chemicals and turning into a serious pollutant. Eventually, the estrogen makes its way out into a lake, pond, or other body of water. There, bacteria convert it back into an active hormone and it starts messing around with all the defenseless tadpoles and roe. Even if you don't care much about the health of wildlife, it's hard to stomach the thought of munching on crackers topped with estrogenic caviar.

Besides this, there's all the packaging involved—you usually get a plastic contraption or package that holds the pills, as well as a cardboard box, receipts, plus instructions and informational pamphlets and whatever bag it all comes in.

On the other hand, not using a contraceptive like the pill means there's a far greater chance of getting pregnant. Although the human race is naturally built to go forth and multiply, there's no denying that having a child directly contributes to the global overpopulation problem, translating into even more limited resources being consumed, not to mention all the garbage and carbon footprints left behind. Furthermore, those who aren't in a committed relationship—or, sometimes, even those who are—must contend with the issue of STDs; taking chances and not using condoms could very well lead to nasty infections, which in turn can mean a lifetime of prescription drugs, which brings us back to the chromosome-warped tadpoles.

Although many hormonally charged hippies recommend using aloe-coated condoms, these are still made with latex and come individually sealed in foil-lined plastic wrappers. If you're staunchly anti-latex and don't want to alter your hormones, however, the only green options remaining when it comes to contraception are either an IUD—a stick of plastic wrapped in copper wire, which gets inserted into the uterus to disrupt its natural state—or what's generally known as the rhythm or calendar method, which involves paying very, very, very careful attention to one's ovulation cycle. Again, though, neither of these protects against STDs. So unfortunately, it seems the only way to be 100 percent safe, 100 percent green, and 100 percent not pregnant is with 100 percent abstinence.

Despite being single and not really having to worry about any of this right now, I was feeling pretty down about these statistics, so I decided to call the one person in my life who was always up.

Meghan.

She went off the pill a couple of years ago, but I had a feeling she wasn't into condoms or IUDs, so I asked her what birth control method she followed.

"Um, well . . ." she said, pausing. "I don't know that you could call it a method, per se."

"Okay, so, what do you do?" I said.

"Well, let's just say I get nervous once a month."

Hmm, this wasn't reassuring. Nervousness only leads to pregnancy tests, which come with just as much packaging—not to mention all the potato chips and magazines that inevitably have to be purchased alongside the pregnancy test in a pathetic attempt to make it seem, to the cashier, as though it was just a casual afterthought. "Yep, just some Frito Lays, the latest issue of *Vanity Fair,* and, oh, why not a little pregnancy test while I'm here? Might as well."

Clearly, Meghan hadn't quite figured this dilemma out either, which was at once reassuring and frustrating. With a feebly protracted sigh, then, I concluded it was a bridge I could cross later, when I got myself an actual boyfriend and something that might be called a sex life.

SEPTEMBER 20, DAY 204

Leave the DO NOT DISTURB sign on the hotel door for duration of my stay

When I found out about my acceptance into a short-story writing program at the Banff Centre, I was excited for three reasons: 1) it would mean I could finally begin some sentences with words like, "Back when I did a short-story writing program at the Banff Centre . . ."; 2) instead of being confronted with mountains of press releases each morning at the office, I'd be confronted with mountains, period; and 3) it would fit perfectly with all my green requirements—the center was in the midst of seeking LEED certification, their cafeteria sourced its food as locally as possible, the residences all had compact fluorescent lights and all-natural toiletries in recyclable packaging, and the sheets and towels were changed only if left on the floor or in the tub.

However, none of these eco-friendly advantages changed the fact that Banff wasn't exactly around the corner—it was in Alberta, which meant yet another flight. More precisely, it meant a 1,682-mile journey to Calgary, followed by a 90-mile bus ride up north. Even if I offset the whole thing, it would still mean that, in this so-called green year of mine, I will have traveled more than some people do in their entire lives.

Still, I couldn't turn down this opportunity for the sake of Mother Nature alone. I love her and all, but this was something I desperately wanted to do. I'd just have to go, appreciate the experience despite the toll it took on the environment, and do my absolute best to leave as small a footprint as possible.

When I arrived and got settled in my room, I noticed the DO NOT DISTURB sign on the outside of my door. I had been told there were still daily housekeeping rounds, which meant vacuuming and garbage bag changes at the very least. So I decided to leave the sign on my door, regardless of whether I was in or out, to ensure there wouldn't be any superfluous cleaning going on.

To my dismay, however, I soon discovered that the housekeeping staff at the Banff Centre is not what one might expect—it's comprised mostly of handsome young men looking for some extra cash to support skiing and snowboarding habits. As I left my room one morning and looked down the hallway, the sight of what looked to be an Abercrombie & Fitch model, decked out in a slim-fitted polo shirt and khakis, pushing a cart full of toilet paper and coffee filters, quite literally made me stop in my tracks.

I wondered how I might change my sign to read DO NOT DISTURB, UNLESS YOU'D LIKE TO COME IN FOR A DRINK.

SEPTEMBER 27, DAY 211

Use a natural saline solution instead of Visine eye drops

My ragweed allergies, which kick in in mid-August and persist until the first frost in December, are pretty intense. When I went to see a specialist who did a pinprick test on me with a needle and a droplet of ragweed oil, he said that in order to be considered officially allergic to the stuff, I needed to have a few millimeters of redness around where he had pricked the skin on my forearm. About ten minutes later, my entire arm—from my wrist up to my armpit—was covered in a horrible, itchy rash.

Now that ragweed season has kicked into high gear, I'm trying to keep my symptoms at bay while also trying to remain as green as possible. This means doing a series of six Pollinex shots instead of taking daily prescription medication, which means I'll be cutting back on packaging and preventing antihistamines from flowing through my system, into the sewers, and eventually into lakes and streams; using this contraption called a Neti-Pot, which naturally cleans out my nasal passages, instead of the headache-inducing steroid nasal sprays; and, as of today, making my own saline eye drops instead of using Visine.

However, after returning home from Banff to what looked more like a demolition zone than my apartment, I'm not even sure I want my sense of sight anymore.

A week ago, I had asked my sister to look after Sophie and my apartment. My housesitting experience with Justin back in the summer was such a success, I figured I'd just print out the same sheet of instructions and leave Emma to take it from there. Because it was for only a week, even if she did forget to water the plants or take out the recycling, it wouldn't be that big of a deal. On top of this, I had just received a fresh CSA delivery, so there was plenty of food. I even plugged the fridge back in to accommodate my sister's need for chilled pinot grigio at all hours of the day. I hadn't heard from her while I was away, so I assumed everything was going well.

Clearly, I assumed wrong. Refusing all the organic, local food I'd left, Emma had instead opted for delivered Thai food, throwing an entire loaf of bread, three whole plums, two apples, a head of lettuce, and a bunch of carrots into the compost bin, without breaking them up or stirring them around into the soil. Looking at this waste would have made any food bank employee weep, but the inside of the fridge was even worse: leftover pizza, half of a takeout ham sandwich, a can of Reddi-wip, a carton of commercial milk, Tetra Paks of imported wine, and so on. Nothing organic, nothing without preservatives—in short, nothing I could eat.

Then I read her note.

It had a title.

"The Deconstruction of a Green Apartment."

"Day 1," it began. "Hair straightening iron got plugged in immediately, fridge dial turned up to make it colder. Left the downstairs light on all night accidentally BUT unplugged the alarm clock as it went off at 4 a.m. and didn't know how to shut it off. Breakfast at Starbucks. Don't know how to cook so will get delivery; this means I avoid using both the stove's energy and my own energy.

"Day 2: Hope the compost bin likes Thai food. I was scared when I looked inside that thing. Borrowed your Marc Jacobs coat this afternoon—borrowing is green, right? Red Bull cans starting to pile up but they're sugar-free so at least I'm not supporting unlocal sugar cane production.

"Day 4: Worms may or may not be dead.

"Day 5: WHO THE HELL INVENTED THE PLUG-IN KETTLE?!? KITCHEN ON FIRE!! MAY DAY! GETTING CAT AND MYSELF OUT OF HERE!"

Apparently, Emma had assumed my kettle was a stove-top one—despite the plug that dangles out quite obviously from its base—and had left it on the element to boil one day in hopes of making instant kimchi while she went back upstairs to do whatever it is she does for hours on end in the bathroom.

The kettle eventually caught fire and started melting, leaving the molten plastic all over the stove and sending smoke up to the ceiling, where it remained scorched.

But this was only the beginning.

In an effort to keep up my green change of "letting it mellow," Emma had also stopped flushing the toilet whenever she went pee. The thing is, as I've mentioned before, my sister drinks only coffee and alcohol and goes number two only every four days or so. This meant that an extremely potent pool of urine was now steeped and festering in the toilet bowl.

Shortly after the kettle incident, Emma went back up to the bathroom, and Sophie, perhaps fed up with the chaos downstairs, decided to follow her. For whatever reason, the cat then proceeded to hop onto the toilet seat and fall in, getting soaked from head to tail in concentrated pee. This, of course, sent her into a panic, so she scrambled out, and began running around the apartment as my sister chased after her, which of course only made her run more, all the while leaving a wet, putrid trail behind her.

At this point, Emma officially gave up and went to her boyfriend's cottage for the weekend.

I flipped over the last page of the note and saw the handwriting switch from her round, bubbly script to my mother's sharp-angled doctor's scribble.

"Came to take over," it said.

Thank god, I thought.

However, I knew from the state of my apartment that something must have gone wrong even when my mother was here because it was only respectably tidy—and she usually keeps every corner of any house perfectly spic and span.

Mom said she managed to patch up the ceiling a bit and properly dispose of all the rotting garbage (Emma couldn't find the chute outside), which was good, but this was a note that I already saw was five pages long, so I held my breath and prepared for yet another un-green chapter in the housesitting saga.

The next page began with a complaint about my lack of cable, which meant my mother had to spend her free time looking through photo albums and—*gasp!*—reading a book. Whether it was the strain of staring at words on a page rather than images on a screen, sheer pop-culture withdrawal, or, most likely, the stress of cleaning up after my sister, by the time she went to bed my mother had the early warning signs of a migraine.

Unable to fall asleep, she decided to get up and make her way downstairs, where there was a bottle of ibuprofen in her purse. Wanting to stick to my green rules and minimize her use of electricity, she opted to leave the lights off while she was up.

Unfortunately, this meant that instead of walking down the stairs, she fell down them.

And fractured her rib.

At this point, my frustration at Emma for disregarding all of my green instructions and nearly burning down the apartment quickly morphed into sympathy for my mother and an overwhelming sense of guilt for dragging both of them into this. What was I thinking? Why was I enforcing this lifestyle on others and nearly killing my own family just to save a few kilowatts of electricity?

"High-dose Advil no help," my mother's note continued.

"Lying on couch all day. Can't move. Need to buy new kettle but housebound. Clafouti croissants not impressive."

This is typical of my mother—broken ribs and shitty croissants are of equal importance.

"Need to bring makeup remover over. Sophie stays upstairs. Why am I here?? Can't find sugar for coffee. Oh no, coffee also needs grinding. Wait, found grinder. Oh, and sugar is on counter. Now coffee has big lumps. At this point, considering wine for breakfast. Oh NO—sugar is SALT!! I need Starbucks.

"Made it outside," the note continued. "Lots of activity on the street. Demo against using cars. Music, dancing, hair-cutting. Feel better but still can't bend or sleep."

A month later, my mother's rib cage had healed and my sister now knew how to properly identify an electric kettle. I had come to realize that, while making one green change every day meant I got to ease into low-impact living, it wasn't a way of life that anyone could just switch over to in an instant. So as more people ask me which changes I think are most important to make, I now pause before answering, taking the time to consider what *their* routine involves on a daily basis, where *they* live, and what *they* value—because while I, personally, was discovering how little I needed a fridge, not everyone can deal with vases full of carrots and room-temperature hemp milk.

SEPTEMBER 29, DAY 213

Eat ice cream in a cone rather than a plastic cup

I thought that writing about Jake Gyllenhaal was the best way to get more hits on my blog—man, was I wrong. Turns out, there's another subject that's guaranteed to get twice as many hits as Jake, and that subject is menstruation. It's totally bizarre—I mean it's not as though the only people procrastinating in the blogosphere are women, let alone women who are very in touch with their bodies and like talking about how in touch with their bodies they are. Or is it?

Either way, I realized this at some point yesterday, when my post about using Lunapads, reusable cloth panty liners, got upward of forty comments, many of which revolved around topics like how to clean them in the sink or a bucket, whether to subsequently empty the bloody water into one's potted plants, and so on, going into enough detail to spoil my homemade lunch.

A funny back story to the Lunapads, though: I'd already bought my own but the company who makes them found my blog and decided to send me a bunch for free. They asked for my address, so I decided to use the one at the *National Post.* The day they arrived, I wasn't in the office, but my editor was. Ben ended up making dinner plans that night with Kelly, one of the reporters who used to work with us, who was in Toronto for a brief visit to see his old friends. Ben decided to bring him some sort of welcome-back present and went rummaging through the Arts & Life swag pile. It was slim pickings, so in desperation he started poking around my desk, searching for any packages of mail that looked like they might be full of promotional goodies. He grabbed the envelope full of Lunapads, took it to the French bistro, and watched in mild horror as Kelly opened it at the table over his duck confit.

Politely declining the gift, Kelly gave it back to Ben, who eventually gave it back to me, and I decided, in turn, to give it away on my blog as a prize to the person who could provide me with an original green change. I was past the halfway mark of my challenge now and running low on ideas, and after all, I didn't see any way in which bribing others to do my creative work for me could be considered environmentally unfriendly.

Instead of turning it into an official contest, however, I just said at the end of my post that whoever could give me a good idea for another green change would win a whole bunch of Lunapads as well as a pair of organic cotton underwear, size medium.

The first to respond was a guy, which I thought was a little odd. It was someone named Mark (not the Mark I knew—I checked the e-mail address), and for a green idea, he wrote the following: "How about asking your male guests to pee in the bathroom sink?

Saves a lot of water and offers a certain pleasurably guilty frisson besides, especially if you have lots of little soaps and tiny towels in the vicinity. If we're going to weather this post-oil decline with any grace at all, we're all going to have to get more comfortable with bodily fluids anyway."

The next comment was from Meghan, who said: "If guests start peeing in your sink, I am never coming over again."

Seriously. That's the grossest idea I've heard yet. Peeing in the shower, maybe, but in the sink? I know urine is sterile but I don't exactly want it splattering onto my mirror and toothbrush.

My sister, Emma, was the next to leave a comment, but she seemed less interested in helping me out with a green change than simply expressing her disgust at my decision to switch to reusable pads. "Am I even related to you?" she said.

Then there were a slew of comments suggesting things I'd already done, which didn't help me at all, as well as some completely outlandish ideas like the one from a woman named Liberty, who said I should consider milling my own grains.

But near the very bottom was a shining green light from reader Laura W, who wrote: "This isn't as hardcore as the other suggestions"—I was liking it already—"but all summer, I've been ordering ice cream in a cone to avoid the dish and spoon."

It was simple. It was cute. It took the prize. In fact, I hadn't realized until now just how ingenious the concept of ice cream in a cone is. If only there were more edible food receptacles in this world. The sandwich is probably the best and most versatile example, of course, along with maybe soup in a bread bowl, but I'm convinced this is only the tip of the iceberg. Surely there must be more avenues to explore here, like, coffee in a muffin, wine in a cheese-glass, or most plausibly, milk in a cookie-cup.

I e-mailed Laura to thank her and shipped off her Lunapads. Only then did it occur to me how odd a transaction this was: you send me a green idea, I send you some menstrual pads.

october

1	Work from home twice a week
2	Hang the bathmat over the tub
3	Smile at a stranger every day
4	No more cling wrap
5	Use wooden clothes hangers; return wire ones to dry cleaner
6	No more swimming pools
7	Pull the shower curtain out so it dries properly
8	Lower the temperature on my water heater
9	Use cash instead of credit cards
10	Reuse tea leaves to remove food odors from my hands
11	Vote for the Green Party
12	Do not pour grease down the drain
13	Borrow and share before buying
14	Let dishes soak
15	Take navy showers
16	Air-dry all laundry
17	No more vitamins and supplements
18	Use scouring pads made of recycled plastic
19	Buy only organic cotton underwear
20	Use the "hit and swish" method
21	Eat seeds and meat from the Halloween pumpkin
22	Abide by the ten-second rule
23	Share my living space with a roommate or guest
24	Use a menstrual cup instead of tampons
25	Bring my own bib to the dentist
26	Set desktop wallpaper to an energy-saving black screen
27	Eat all fruit down to the core
28	Play eco-friendly board games
29	Chop food manually before using the food processor
30	No more hand sanitizer
31	Wear a minimal/reused Halloween costume

OCTOBER 1, DAY 215

Work from home twice a week

I've convinced my editor to let me work from home at least twice a week, partly because telecommuting is green and partly because my journalistic creativity flows better when I'm in sweatpants on a couch. The problem is, I'm not so sure I want to be at home right now because a few dozen unwanted guests have just turned up in my kitchen and living room—the compost worms, and their nasty relatives, the maggot family.

Let's take this back for a second, to when I first returned from Banff to find my apartment strewn with half-empty Styrofoam containers of pad thai, a burnt ceiling, and a series of notes from my mother and sister detailing how their housesitting efforts had gone horribly awry. I had spent the remainder of that evening cleaning out the fridge, sifting through the garbage for recyclables, and getting the piss smell out of my cat's fur. Stupidly, however, when I unplugged the fridge again, I neglected to check whether Emma had left anything in the freezer. As it turned out, she had: a microwave dinner of Michelina's lasagna. And Michelina's lasagna, when it sits at room temperature in a dank, stagnant environment with no ventilation whatsoever, does not fare well.

This afternoon, then, when I opened the freezer door to let some air in, I smelled something so putrid, I gagged, spun around, and bolted for the kitchen sink in a dry-heaving revolt. Creepy-crawly larvae had colonized the entire top half of the appliance, and it reeked of dead flesh and wet cardboard.

After collecting myself, I decided the most sensible thing to do was simply keep the door shut and ignore the infestation. In another four months, I could turn the appliance back on, crank the dials up, and at least freeze the suckers to death. Then I'd make my new boyfriend get rid of the mess.

Standing there in the kitchen, though, I was reminded of yet another potentially revolting green chore that needed doing immediately: the weather had been getting colder, and my compost bin would have to be relocated from my balcony into my living room to ensure the worms stayed alive. There was a corner between the bookshelf and the dining table where the bin would just about fit, and although this makeshift plywood box wasn't exactly what you'd call a design feature, I figured it would at least make for a decent conversation piece. (Further to this, I was hoping that maybe sharing my apartment with a few dozen worms would make me eligible for a tax rebate. Could I claim some sort of interspecies common-law status after six months?)

Either way, it had to come in.

When I'd first constructed the thing, I had my friend Sam there to help move it out before filling it up with soil, food scraps, and worms. Now it was three-quarters full and there was no one here to help. Nonetheless, I was sure it couldn't be that difficult—it was still just a box that had to go from one place to another, no more than a few feet in distance. Stepping outside, then, I grabbed the unit from both sides and yanked hard, so it slid toward me. I could feel tiny slivers of plywood threatening my forearms, and the front banged hard into my gut. I decided to rotate it so it faced the sliding doors that opened into the living room and lined up better. This would be the hard part, though: getting it up and over the balcony door ledge, then through to the living room. It was a tight fit.

Although the bin could be pushed and pulled, up and over, with a few strong heaves, it wasn't so simple trying to lift it because there was nothing to grab onto other than flat surfaces—the lid ran flush with the sides, which didn't have any handles or even a ridge for my fingers to grasp. So in the end, I just draped myself right across the whole thing, grabbed the back corners, and, using every muscle that physiotherapists tell you not to use, yanked the box toward me with a strained yelp.

It almost worked. I got it halfway through the door before it got stuck. At this point I lost my grip and fell backwards.

This wasn't such a problem, the falling, but at the same time that I hit the ground, the bottom drawer of the bin dropped down and rolled away while the chicken wire holding the compost ripped off the inner wall and collapsed, effectively sending a stream of slimy cabbage, frantic worms, and pungent soil straight onto my living room floor.

I couldn't believe it.

It was like a scene from the green twilight zone.

Sitting there, staring at this pile of mulch at the foot of my couch, I was on the verge of one of two reactions: I could either feel sorry for myself and cry, or I could appreciate the absurdity of it and laugh. This decision wasn't easy—if there's anything I excel at, it's self-pity, and crying was looking like a pretty good option—but after a few seconds of sitting cross-legged with my head in my hands, watching as my worms wriggled their way across the floor, trying to find whatever victual remnant they were last working on, I finally chose laughter and grabbed a dustpan.

OCTOBER 4, DAY 218

No more cling wrap

Compost junkies. That's what they are.

Janet, the leader of my Toronto Environmental Volunteers group, had warned me about these guys at one of our group training sessions, where we were learning everything we'd need to know about the upcoming series of environment days, which were held every week at a different location across the city. This was where residents could drop off hazardous waste for proper disposal as well as number 6 polystyrene, plastic bags, and cling wrap—which I'd given up today as my green change—for recycling. They could also pick up new recycle bins, compost units, and water efficiency kits; donate any clothes, eyeglasses, or furniture to Goodwill; drop off food at the food bank; or stop at one of the information tents to learn about other municipally funded programs.

But the reason these environment days were so popular actually had nothing to do with the cork recycling, battery drop-offs, and information booths—no, it had to do with mulch. A huge, eight-foot-high, twelve-ton pile of mulch.

"You have to stand back from the pile to make sure you don't get hit with a shovel," advised Janet. "People really attack it. Sometimes the whole lot's gone in twenty minutes. We once had two cops manning the pile and when we said go to the crowd, they all rushed in and the police just turned and ducked."

Apparently, this compost is the best fertilizer around, and, more importantly, it's free. A mix of leaf and food waste, it spends about a year in an anaerobic digester out in the suburbs until it's ready for use. Each week, some of this compost is loaded into a dump truck and driven to wherever the environment day is being hosted, where it's tipped out onto the grass or asphalt. If the ward's councilor has a big enough budget that year and really wants to impress her voters, she'll sometimes even spring for a second truckful.

Prior to the meeting, I'd spoken to the manager of Toronto's Solid Waste Services, Geoff Rathbone, for a story I was writing at work about composting. He's helped to run the environment days for over a decade now and has watched the popularity of the compost pile grow over the years, from a curious crowd into a near-maniacal, frenzied mob.

"Over thirty-two thousand people showed up last year in total," he said. "The compost people tend to show up about an hour before it gets dumped, to get a good spot. We usually have some crowd control issue . . . The pile gets completely surrounded, and some people even climb on top to get a better spot. Within a very short period most of it's completely diminished, but the real keeners will stick around to sweep up any remnants left over. Especially if we're hosting it in a parking lot, it'll get in the cracks of the asphalt."

Sweeping flecks of mulch out of cracks in the pavement?

I had to see this firsthand.

And so this is how I found myself standing in the pouring rain today, searching for Janet, whom I predicted would be wearing a Tilley hat (and she was, one with buttons). I waved hello and she promptly came strutting toward me with a smile so genuine its warmth made up for the lack of sun. She waved off my apology for being late and led me over to one of the tents where some of the other volunteers had congregated. After a quick round of introductions, she suggested I go hang out with the solid waste crew, who take care of all the Styrofoam and plastic drop-offs. I was handed a pair of flower-print gardening gloves and a green TEV polo shirt and told to watch and learn.

It didn't take much watching to figure it out: there were two bins for cling wrap and plastic bags—any type was fine, as long as it didn't have a drawstring closure (thanks, Gap)—and two bins for polystyrene and plastic takeout containers. Only Styrofoam and plastics with the number 6 printed on them were acceptable, which meant no packing peanuts or harder, unbreakable stuff. The test was always whether the material could be ripped; if not, it was garbage. Once the bins were full, we took them in bags over to the truck and let the workers toss them in.

Today's solid waste team was comprised of three guys: Brian, Luke, and Jason. Brian was thirty but the others were both twenty-something; all were part-time employees of the city of Toronto and all of them, at least in my opinion, were cute.

Well, hey, I thought to myself. Of all the places to look for date-worthy men, the solid waste sorting department is coming up on top. Maybe after this, I should check out the sewage plant and pest control division.

After about an hour, when the cold came in to accompany the rain, the guys decided to wander over to the animal services tent, where they were handing out brochures about pet adoption and answering whatever questions people might have about raccoons or squirrels in their backyard. They also had information about their euthanasia services. I was half-impressed and half-disturbed to find out there was a giant refrigerator at animal services headquarters where they stored all the dead animals—not just cats and dogs, but all sorts of roadkill, too—before they're incinerated.

"They're in garbage bags to keep the fridge clean but we can't actually burn the bags," said the young, petite brunette manning the booth, "so it's kind of annoying 'cause then we have to take them back out of the bags before throwing them in the pile to be burned."

I stared at her in horror. She was clearly enjoying this.

"The worst part, though," she said, "is in the summer, on really hot days, when you're scraping dead animals off the street and there are all these maggots, and then the power cuts out so the fridge starts to smell."

My stomach lurched right through my face.

"Yeah," she went on, "honestly, you couldn't pay me to go into work those days."

The question of how she could simply not go into work even though she was a full-time employee just flew through my brain. The only thoughts I had were of rotting squirrel carcasses, blood matted on black fur, flies feeding on eyeball juice.

"Who wants cookies?" came a voice from behind me.

It was another one of the lead volunteers, carrying a plastic container of chocolate chip cookies from a nearby convenience store. Everyone reached in to grab one, and, while I'd sworn off any chocolate that wasn't fair-trade, I also hated to come across as a girl who watches her weight or cares about whether something is gluten-free, dairy-free, or whatever else, so I dug in. It was stale.

"Hey, I can't believe you bought food in a disposable container on environment day," I said to the volunteer, in as lighthearted and nonaccusatory a way as possible.

"Yeah, but I can recycle it right over there," she countered, gesturing toward our unmanned booth. Good point. It was a number 6 plastic and it ripped easily.

"It's still a little ironic though, no?" I pushed.

"There's a lot of irony here," came another voice to my left. It was Janet. She'd wandered over to see what we were up to and was holding a Starbucks cup in her hand.

"Hey, you too!" I said. "You can't recycle that, can you?"

"No," she replied, "but it's caffeine, and it's free, so I'll take it."

The more I actually talked to Janet, the more I realized there was an edge to her. And in the green world, or at least in my green world, this is gold.

"So what were you saying about irony?" I asked. "I mean, I have noticed a lot of coffee cups around here, not to mention all this bottled water they're giving out."

What was bizarre in particular, I thought, was that if you showed up at environment day and wanted to recycle basic things like plastic bottles or newspaper, you couldn't. We accepted only what wasn't taken by the garbage and recycling trucks every week.

"Well, you know," said Janet, "it's just funny how all these people make a point of holding on to their Styrofoam, old tires, paint cans, and everything, but then they pack it all into the car, drive here, and then leave the engine idling for ten minutes while they go around putting everything in its place."

I nodded, looking behind me at the rows of cars. But then again, how could you transport all that stuff on a bike? It was a bit of a Catch-22, really.

"Then there are all the pamphlets and brochures and schedules we hand out here, which have all the information people need to know, but then more paper is used," Janet added.

Again I nodded, understanding and yet having no idea what the solution might be. You couldn't exactly have an electronic environment day.

By now, the rain and the cold had been joined by gusty winds. Janet reached into a box underneath one of the display tables, pulled out a plastic blue poncho with the TEV logo on it, and handed it to me.

"Here," she said. "It's yours. It'll keep you dry."

"But I already have a windbreaker on."

"Yeah, but you might as well take another one. It's free."

Great, I thought, putting it underneath one of the benches behind me so I could return it, intact and unused, at the end of my shift. More free crap, more irony, and now even more doubt about just how effective this volunteering effort really was.

OCTOBER 8, DAY 222

Lower the temperature on my water heater

Over the past few years, as I've dealt with various leaks, faulty light switches, broken heating units, and so on, I've gradually learned more and more about the things that make me clean when I'm dirty, warm me up when I'm cold, cool me down when I'm hot, help me eat and drink, and let me see when it's dark.

I truly believe the greatest gap in our modern education system lies in one of the most basic areas: houses and how they work. What's the point of learning the speed of train A coming from X as it approaches train B coming from Y that's traveling at two-thirds the speed but carrying half as many people if, meanwhile, you have no clue how your bedroom is being heated and what happens to everything you flush down the toilet?

Take my water heater, for instance. I'd made 221 changes to my life, all with the goal of reducing my carbon footprint—nineteen of which were related to water in some way—and all this time, I had an energy-sucking monster hiding in my storage closet, using lots of electricity to keep a big tank full of water far hotter than it would ever need to be.

I'd just heard about this water heater thing, so I went to check on mine and get better acquainted. The temperature was set at 140 degrees Fahrenheit. I found conflicting information online about the temperature to which it could safely be lowered, but I decided to go with 110, which meant the water still got hot—just below what the hottest hot tub can be cranked to, apparently—but far from scalding.

The only risk associated with this, according to my research, is Legionnaires' disease.

A fun fact about Legionnaires' disease: it was discovered and its name coined in 1976, after delegates at an American Legion convention in Philadelphia got sick from a type of bacteria usually found in aquatic or wet environments, which often leads to pneumonia. Scientists decided to name it *Legionella pneumophila.* It's not exactly clear what sort of water-themed activities the Legion members were getting into that week back in '76, but either way, the illness turned out to be a potentially fatal one, with a substantial 5 to 15 percent of the cases resulting in death.

Some commenters on my blog warned me of this when I posted about the change, but upon further research, I found this statement from the U.S. Department of Labor: "Evidence indicates that smaller water systems such as those used in homes are not as likely to be infected with [Legionnaires' disease] as larger systems in workplaces and public buildings."

So if I keel over and die in a few weeks, I'll blame the Yanks.

OCTOBER 17, DAY 231

No more vitamins and supplements

The best comment on my blog today came from my dad, who wrote, "The only supplement I take is a nice single malt before bedtime—works wonders."

OCTOBER 18, DAY 232

Use scouring pads made of recycled plastic

I have a file on my computer desktop called "The List." It's where I keep track of all the ideas for this challenge—any suggestions my family, friends, and readers give me, or even just links to articles and blogs that have featured something I think I could adapt to my own lifestyle. I used to keep this list on an actual eight-by-ten-inch sheet of paper, but it started turning into a complicated matrix of flow charts, crossed-out and circled words, arrows, lines, and multicolored ink, so eventually I tossed that in the recycling bin and switched over to an electronic version.

Every time I use one of the ideas on The List, I highlight it and click delete. During my more organized weeks, I'll have a running tally of about ten to fifteen things and will choose what I'm in the mood for doing. But most of my weeks are not at all organized, which means The List hovers around three or four things to choose from. A really bad week means I'm stuck between unplugging yet another appliance or inhaling my own exhales.

Some things aren't nearly as drastic but remain on The List week after week, and often month after month, simply because I'm dreading making the change. One of these things is giving up my vacuum cleaner and using a broom and dustpan instead. To most people, this would seem like an easy change, especially seeing as my floors are all hardwood and tile. But I just have this inexplicable love of vacuuming—it's my favorite of all the chores. I love the whole concept of sucking up dirt rather than sweeping it around; I love the sound and the feel of it, the way it's like some powerful weapon wiping out all traces of dust, hair, and crumbs. It's the difference between clean and spotless, and, well, to paraphrase Charlton Heston: you'll have to pry my vacuum cleaner from my cold, dead hands before you get me to pick up a broom.

Along with giving up the vacuum, though, there's another thing that's been sitting on The List for ages now: the eco-scrubby thing. That's exactly what I've written, too: "the eco-scrubby thing." But it's not still on my list because I'm hesitating to make the change—it's there because I wrote it down months ago and forgot what the heck it was.

I bought it impulsively a few months ago at this kitchen store downtown, and the cashier told me it was environmentally friendly. It's basically a rainbow-colored scouring pad. Finally, today, I decided to call up Tony, the manager of the store, and ask him once again what qualified this thing as eco.

Turns out, it's made from recycled plastic. It was also manufactured somewhat locally, in Winnipeg, Manitoba.

All right, so it's not exactly going to put an end to global warming and save the polar bears anytime soon. Still, it was good enough for me to leave by the sink in case of sticky-old-scrambled-free-range-egg emergencies and call it another day.

But now my list was nearly down to a blank page, which put me on edge. I stared at the ominous words *broom instead of vacuum* and wished I could just sweep that whole idea under the rug. Eventually, though, I knew I'd have to suck it up and pick up a damn broom because it's the right thing to do, and environmentalism isn't always about doing what you want. In fact, something tells me there might not be a single environmentalist who enjoys scrubbing the toilet with baking soda by hand—but there are plenty of them who do it on a regular basis.

With this in mind, I decided to solicit some empathy and ask my readers at the end of today's post to tell me the number one green thing they always make a point of doing, no matter how much they hate doing it.

"Washing out Ziploc bags and reusing them," said Teaspoon. Hellcat13 added that she hated bringing leftover food back from the office to empty into her compost bin at home.

Then there was Chile, who wrote, "Environmentally friendly pest control doesn't work nearly as well as chemically bombing the heck out of my yard." A fair observation.

All of these were relatively minor, understandable gripes. But then came a comment from someone named Bryan that I am very happy I did not happen to read while eating my lunch: "I use cloth nappies (diapers) and cloth liners," he wrote. "I wash by hand. And, our poo gets used as humanure. This means that to get the cloth liners clean, I must first soak them, creating a bucket of poo-soup that after a mere 12 or so hours threatens to crawl all by itself to the 'worm farm' where our bucket-dunny also gets emptied. The really gross thing is getting the cloth liners out of the soup before emptying the broth, and releasing the odor that has been hitherto sealed beneath the layer of scum—the fat of the soup."

As I would have said back in grade school: sickatating.

OCTOBER 24, DAY 238

Use a menstrual cup instead of tampons

Of all the suggestions people have offered me throughout this challenge, there's been one that has continued to come up over and over again. It has nothing to do with recycling, or tote bags, or going vegetarian. It involves menstruation and this thing called the Diva Cup, a reusable silicone alternative to tampons—there's also the Keeper, a similar contraption made of rubber—invented by a mother-daughter team from Kitchener, Ontario, about five years ago. (According to the online Museum of Menstruation, different varieties of menstrual cups have been around since the 1930s. And yes, there actually is an online Museum of Menstruation.)

While at first the idea of an insertable cup sounded completely bizarre, when Meghan said she had made the switch, I decided to give it some real consideration.

On the phone one night, she explained that her motivation to try the cup stemmed from learning how tampons were made.

"Let me e-mail you the essay I wrote for my nutrition and the environment class," she said, and so I waited for her to send it while still on the phone. When I finally opened the attachment, I couldn't help but laugh at the title.

"*Tampax Tampons: Toxic Death Sticks?*" I said. "A little on the melodramatic side, Meg."

"I wanted to make sure I got my point across," she replied.

After we said our goodbyes and hung up, I started reading.

The essay itself wasn't as over the top as the title, and began with the rather sound argument that women shouldn't take for granted any product they put into the most intimate part of their body on a regular basis—a product, in this case, with the sole purpose of absorbing blood and temporarily blocking its natural flow. As well, considering most women have a period that's five days long, occurring every four weeks for over thirty-five years, this translates into 11,400 tampons per woman, which don't just clog up sewer systems but leave traces of dioxin—a potential carcinogen in the chlorine that's used to bleach the cotton—lingering in this delicate region.

Meghan continued for over ten pages about the pesticides used on cotton fields and the hundreds of different chemicals involved in the conversion of wood pulp to rayon, which Tampax uses for absorbency. It all sounded very important, but words like *dioxin* tend to bore me, so I just called her up again and asked some more pressing questions, such as:

1) How exactly do you get a whole cup in there?
2) Can you feel it? At all?
3) What happens if you need to empty it and you're stuck at the office? Or at a restaurant? Or at the mall? How do you rinse it without anyone seeing?

"You just fold it up and stick it in," she said. "You can't feel it once it's up there, and you only need to empty it once a day, so just do it in the morning or before you go to bed.

"Seriously," she added. "It's sooo much better."

I winced with skepticism. But Meghan's rhetoric was convincing, so I decided to at least try it. The only problem now was that, after going off the pill last month, the inner workings of my body were all a mess. My ovaries had slept through their regularly scheduled appointment with my uterus, my progesterone didn't know whether it was coming or going, and so the Diva Cup I'd purchased was now languishing untouched in its fuchsia drawstring pouch under the bathroom sink.

But finally, out of nowhere, things kicked into gear. Unfortunately, the bleeding happened in the middle of the night with nary a warning cramp, so after I'd hurriedly scampered into the bathroom and flipped on the lights, the whole insertion process was conducted in a fumbling, somnambulant daze. But I eventually got it in, or at least got it far enough that it would last me for a few hours, and made my way back to bed. When I woke up that morning and went to check on things, however, there was a slight complication: the cup was stuck.

When I'd read Crunchy Chicken's blog post about the Diva Cup, I noticed many of the women who'd left comments said they needed to trim the little stem at the end, but in my case I could have done with a much longer one, or at least something—anything—to grab on to. But I remained calm, pulled up my pants, and hobbled over to my computer.

"HELP DIVA CUP STUCK," I typed into Google, and in less than five seconds the troubleshooting section of the Diva Cup website was up on my screen. There were instructions for precisely what to do in this scenario—instructions that involved squatting and a so-called bearing down action, which I could only just improvise, really—and luckily, after a few seconds, the cup was out and the problem was solved.

It took only another day or two to get the hang of it, and I was quickly beginning to realize why so many women had been pleading with me to try this thing. Regardless of the health benefits, the Diva Cup was actually easier to use and more comfortable than tampons, created no waste whatsoever, and meant that I wouldn't have to spend money on other menstrual products every month. Nothing about the process was gross or made me cringe, and the cup never leaked or needed changing during the day.

I was officially a convert.

Checking the comments later on tonight, though, I got a little worried. Someone by the name of Peregrine had written: "To those concerned about physical activity with the Diva Cup, I've been swimming, dancing and biking with mine and had no problems. However, doing yoga one morning, I was in a downward dog, lifted one of my legs up in the air and started pushing it back in the direction of my head, and heard that signature pop of the seal breaking. Needless to say, I was extremely grateful that I was home and could run to the bathroom to avert disaster . . . My hips are really loosely jointed and some of the poses I do apparently pull my cervix out of its normal cylindrical shape, thereby allowing the cup to leak. So you are warned. But I have to say, I still love the thing unconditionally."

Signature pop? I did not like the sound of that.

OCTOBER 25, DAY 239

Bring my own bib to the dentist

"*Buim, Dr. Daniel. 416-921-. . .*" said the call display on my phone at work today. It was my dentist's office. But I'd just been to my dentist. Why were they calling? They already knew about whatever cavities or tartar buildup I had. Had I left my handkerchief-turned-bib there by accident? Were they upset with me for declining the fluoride treatment and complimentary toothbrush? Or was this like when the doctor's office called with bad news? Did I have tooth cancer? Is there such a thing as tooth cancer?

I let it go to voice mail.

When the red light started blinking to let me know I had a message waiting, I dialed the appropriate numbers to check it.

"Hello, hi," said a voice, in a hushed tone. "Um, this is Daniel Buim's office calling. I'm actually the secretary here, and, um, I couldn't help but overhear your conversation, you know, with the hygienist the other day, after your appointment, when you were leaving."

She really spoke in commas, this woman.

"You were, I think, talking about this thing, I think it was called the Diva Cup? I just wanted to know, if you wouldn't mind, could I ask you a couple questions about it?"

I chuckled and immediately phoned her back. In the end, she'd just wanted to know the basics of how it worked and where she could buy one, and it was altogether a sort of whispered and girly conversation. Maybe it was just my relief at not being told I have tooth cancer, but I got a real thrill out of knowing I might have just prevented another few hundred bleached wads of cotton from clogging up the sewer systems.

OCTOBER 28, DAY 242

Play eco-friendly board games

Manse: a big house or residence of a clergyman.

I will never again forget what a manse is. Not after today.

I've discovered this online game called Free Rice. For vocab nerds like me, it's addictive, far more so than even Scrabble—it presents a word with four possible definitions below, and after every definition you get right, ten grains of rice are donated through the UN to help conquer world hunger. I got a top score of forty, which means I gave over one thousand grains of rice, but if it weren't for not knowing what a stupid manse was, I probably could have scored forty-one. I wonder if anyone's ever built an off-the-grid manse?

OCTOBER 31, DAY 245

Wear a minimal/reused Halloween costume

Meghan's family, the Telpners, are the type of people who take dressing up very seriously. Meghan originally went to Ryerson University to study fashion, where she won the faculty gold medal for top marks and achievement, and can also make her own clothes and jewelry. Her father, Ron, has over one hundred pairs of designer shoes and a diamond embedded in his tooth (the top right incisor, I believe). Her mother, Patsy, and brother Michael are less concerned with the fashion industry, but nonetheless seem to appreciate any opportunity to get into costume. So when Halloween rolls around, the entire neighborhood knows to stop by their place because they not only give out lots of candy, they go all out with the decorations, the music, and of course the outfits.

This year, as I approached their house, I began shaking my head in unsurprised surprise. They'd gone with some sort of retro Polynesian zoo theme. There was a bunch of yellow tape cordoning off the front lawn, which was scattered with pretend human remains. Suddenly I heard a loud growling noise and turned to see a giant ape appear from behind the tree, banging his chest and stomping over toward me. Michael was inside, I figured out. Soon after, Meg and her sister-in-law stepped out of the house in psychedelic go-go-girl dresses, complete with white pleather knee-high boots and enormous sunglasses. And at the front, by the sidewalk, stood her mother and father, wearing safari hats and suits, manning the candy station and makeshift tiki bar.

I walked up to the bar, of course.

"What can I get you?" bellowed Ron, after pouring one of the parents a double-something on the rocks. Did he have a liquor license? It was definitely real booze, and the parents all seemed overwhelmingly grateful to have access to it.

"I don't know," I said. "I kind of feel like I should get a costume on before I get my drink on."

Standing there in my jeans, zip-up sweatshirt, and running shoes, I felt completely inadequate. I'd pledged as my green change for the day to wear only a costume that was assembled from used or recycled material, but I'd been too busy—or really, too lazy—to actually put anything together, and thought if I desperately needed something I could just stick some leaves in my hair, some bark on my chest, and say I was a tree-hugger.

This clearly wouldn't be enough, though, and hardly fit the gory, safari (or rather, gory, safari disco) theme of the evening. I turned to Meghan's mother, Patsy, and asked if she had any other outfits or masks, anything at all.

"Hmm," she said, walking inside to the hallway table, where she picked up something that might be described by certain cannibals as a bracelet—a plastic severed hand and forearm that looked like it had been chewed off at the elbow and which hooked around the wearer's wrist so as to suggest he or she had done the chewing. She handed it to me, gnawed-off end first, and said I could probably use it in some way when giving out candy.

I did as she said but it wasn't long before I realized how pathetic I looked. All the kids coming up to the house were obviously more interested in the crazed gorilla and trippy go-go dancers, the parents were more interested in the free booze, and not even a bag full of free candy could entice anyone to come over to me. Thinking back to it, I probably looked like some too-cool-for-Halloween hipster who accidentally stumbled into north Toronto after a crippling bout of ennui . . . and a pit stop at Hannibal Lecter's.

Claiming that my lack of a costume was environmentally friendly wasn't getting me anywhere tonight. It occurred to me, then, that while being proactively green—as in, making the effort to sew a costume together from old clothes and handing out organic candy—could be justified, being passively green and not doing anything whatsoever was just lame.

I fished around in the candy bag for something good, something sweet and sticky to chew out my shame and maybe get a bit of a sugar rush—even holistic nutritionist Meghan was indulging in a Kit-Kat bar, after all—but everything was individually packaged in plastic or foil or made with non-fair-trade chocolate.

Karma was totally kicking my ass.

november

1	No more Post-Its
2	Clean dryer lint traps and filters on a regular basis
3	No more Dustbuster
4	Use aloe-coated Beyond Seven condoms
5	No more toothpicks
6	Buy and share big sturdy golf umbrellas
7	Ask others to do green things I can't do
8	Green my nervous breakdown with tea and meditation instead of drugs
9	Enforce a daily quiet time
10	Reuse old socks as cleaning rags
11	Switch to all-natural carpet cleaner
12	Keep my bathroom fan off
13	Dispose of electronic waste properly
14	Back up data with reusable USB drive instead of CDs
15	Support charities without wearing wristbands or ribbons
16	Reuse old computer discs as coasters
17	Drive the speed limit
18	Drink straight from the bottle
19	Do not wash fruits and vegetables (unless they're really dirty)
20	Knit my own scarves and mittens
21	No more new hair accessories
22	Polish silverware with baking soda
23	Give used magazines to the doctor's office
24	Buy and refurbish used furniture
25	Learn to sew and mend clothes
26	Reuse my artificial Christmas tree
27	Ask for only green gifts
28	Keep address book electronically
29	Use coconut oil to polish my shoes
30	Make soup broth from scratch

NOVEMBER 8, DAY 253

Green my nervous breakdown with tea and meditation instead of drugs

The only way to relate the nerve-shattering saga of the past forty-eight hours is, I think, with a timeline. So here:

Tuesday, 10 a.m.—My mother, who is obsessed with checking the local real estate listings, tells me about a semidetached brick house just around the corner that's been completely renovated with three floors and a garden and is going for a ridiculously low price because it doesn't have parking. I tell her I can't afford a house and she tells me I should at least take a peek at it during the open house. Because I'm working from home and because I always succumb to matriarchal pressure despite my ongoing attempts not to, I stop by after refilling my almond butter jar at the health food store down the street. Immediately, I fall in love. I must have this house.

Noon—By now, the agent has been contacted; a barrage of frantic cell phone calls, text messages, and e-mails have been exchanged; an offer has been drafted.

4 p.m.—All attempts at working on things related to my actual job have gone out the window as I fax the offer, counteroffers, and counter-counteroffers to the current owners, all the while keeping my agent on the line and negotiating prices with the bank, i.e., the Bank of Mom and Dad.

6:30 p.m.—More important than buying a house at this point is the fact that, in less than an hour, I'm supposed to go on a date with a guy I've been pursuing ever-so-strategically for the past two months and who I'm convinced is my soulmate. He's a film critic at another local paper, and although he's divorced with two kids, he is also cute beyond all comprehension—truly, I believe the answer to world peace may be found within his dimples—not to mention smart and funny. And to top it all off, he has impeccable taste in pop culture, with sarcasm and self-deprecation in all the right places. We have plans to put on our skinny jeans and sit on his brand-new couch, then indulge in some HBO, ice cream, and other couch-oriented pursuits. The problem is that I might have to buy a house before I can buy any locally churned organic ice cream, and so this in turn means I'm going to be late and frantic and starving and dizzy and most likely perspiring from multiple parts of my body by the time I finally get to his place.

8:30 p.m.—I am the fucking owner of a fucking house.

8:31 p.m.—My agent and I down two hefty glasses of wine.

9:05 p.m.—I am standing outside my crush's apartment, without ice cream but with a six-pack of locally distilled beer, which is all I could manage to find between buying a house and getting on the streetcar. I also think I might puke, and the natural deodorant is really not pulling its weight.

Wednesday, 3:30 a.m.—After six hours, I arrive home from one of the most successful and yet unfulfilling dates I've ever had. Food and drink were perfect, conversation was perfect, venue was perfect, and it really did surprise us when we looked at the clock and saw that it was three in the morning. But nothing happened. No moves were made, the flirtation never went beyond an elbow click or occasionally leaning over one another to retrieve a beer. When the cab showed up at his door, he pecked me on the cheek and walked back upstairs. I felt a bit disappointed, but managed to fall asleep by reassuring myself that my life, at least on paper, was good: I'd just bought a house and gone on a date, and neither of these things went wrong in any way.

8:00 a.m.—Almost five hours later, I wake up, scramble into a pair of stale jeans and a pockmarked cardigan, whip my tangled hair into a ponytail, and bike as fast as possible to the penthouse suite at the SoHo Met Hotel, where I'm about to be not-so-casually late for an interview with British celebrity chef Nigella Lawson. When I arrive, I shake her cool, delicate hand and try not to stare at her legendary bosom. Then I ask her about the travesties of plane food, various cookbook clichés, gastronomic trends, and whatever else I can throw into the allotted twenty-minute block.

11:30 a.m.—Get home, but realize I've forgotten to go to the bank to pick up the check for the real estate agent.

11:40 a.m.—Arrive at the bank. Manager is on lunch. I'm told to come back later, so I head home.

11:50 a.m.—Transcribe the Nigella interview and begin writing the story but can't stop thinking about last night's date so I quickly send my crush an inquisitive yet easy-breezy text message, trying to suss out his feelings.

Noon—Get the lead of my story down, then receive a text back saying he had a great night and was impressed I'd been able to keep him up that late despite his oncoming sore throat and cold. But it didn't exactly conclude with "so let's do that again," and so I continue to feel nonplussed.

1:00 p.m.—Stare at the same paragraph on my computer for an hour, and do nothing other than change a comma. The story is due at 2 p.m. I decide to do what any responsible journalist would—I check my Facebook account.

1:05 p.m.—Inbox (1). Click on a message from the boy. He's being more forthright now, and again it's bittersweet. But mostly bitter. He says he likes me and wants to see where our friendship progresses but is not nearly ready for the dating scene in his post-divorce, post-intense-rebound-fling, post-career-crisis state, and needs some enforced singledom for a while. It's the nicest rejection letter I've ever received, which really just makes it worse, and now I definitely can't concentrate on my article.

3:30 p.m.—I forward the e-mail exchange to Ian and implore him for emotional support.

"Wow, that's too bad," he deadpans. "He seems like a really nice guy. I can tell from his message. I mean, it's clearly just not the right time."

"Yeah, I guess," I say quietly.

"Hmm . . . Sorry about that."

This is not helping. I hang up.

5:25 p.m.—Now three and a half hours past deadline, I conjure up every ounce of mental strength, repress all emotional sentiments, imagine my editor firing my sorry ass, buckle down, and pound out the rest of the Nigella story, then send it off to Ben with a note of apology, sit back, and realize the bank is about to close.

5:35 p.m.—Pedal as fast as possible back to the bank and get yelled at by a pedestrian for being on the sidewalk, which I was using only because there was construction on the road, but now I feel like a bad cyclist anyway. Finally arrive at the bank. It's closed.

5:45 p.m.—At home again and have all of ten minutes to get to the Canadian Opera Company to do yet another story, this time about a group of Toronto men who are either bald or "willing to be bald"—they're auditioning to be extras in the upcoming production of Leos Janacek's *From the House of the Dead*, which for some reason requires a lot of bald dudes and for some other reason demands a feature article in the *Post*. I have to interview them, fortunately not on deadline, but there's no way I can make it there in time on my bicycle, so I get back onto the computer, bring up the Zipcar homepage, and book a Honda Civic in the lot outside my building, run down, hop into the car, and floor it to the East End.

6:10 p.m.—I'm late, but the publicist is forgiving, as publicists should be. I conduct interview after interview, about everything from the stigma of balding to the future of Canadian opera, and because all the extras are very eager to get into the paper, I unfortunately don't make it out of there until a couple of hours later.

8:30 p.m.—Arrive home, remember the agent is coming over any second now to talk about selling the apartment, and start running around doing as much cleaning as I can.

8:45 p.m.—Realize I haven't actually consumed anything resembling food today, so I stick a piece of stale bread into the toaster, spread some organic garlic butter on it, and shove it in my mouth just as the agent buzzes me from downstairs.

8:50 p.m.—Spend an hour or so going through her market analysis and what the selling process will involve, then follow her as she walks around pointing at all the things that "have to go" before the open house, which she's scheduling for this weekend. Among these things are my compost bin, my cat, and my bike—arguably my three most prized possessions at this point in time. She then asks me to write down all the stuff I'll need to clean, like the stairwell, and all the things I'll need to pick up at IKEA tomorrow (I'm going to IKEA tomorrow?), in order to "fluff" the place. This includes monochrome pillows, ornamental flowers and/or apples, as well as "those little white curvy vases and maybe a bookend or two, like the Eiffel Tower–shaped ones."

11:15 p.m.—By the time she leaves and I've finished checking my e-mail and sorting out everything to do with work, it's a quarter past eleven. I have no idea what I'm going to do for my green change tomorrow, there's nothing left on The List to help me, and now I have to sell my apartment and move into a house, all the while trying not to get fired from my job and also trying not to think about the fact that I don't have anyone in my life to hold me together when I fall apart. What the hell was the point of moving into a three-story house if there was no one to share it with? What was the point of any of this—the composting, the fridge unplugging, the recycled toilet paper—when I was doing it all alone? The more my heart broke on the inside, the less the outside world mattered; the planet, at this point, could piss off and die.

Aaaand . . . cue nervous breakdown.

I cried, and I cried, and I cried. Then the crying turned into shaking, nose-running, and sobbing, until eventually I began hyperventilating. By the time the convulsions were in full force, I realized I'd have to calm myself down. Immediately, my thoughts flew upstairs, to the bag full of Advil, Gravol, and other medication under the bathroom sink. In there, I had a couple of expired Ativan, an antianxiety drug I used to take on planes, back when I had a fear of flying.

They always worked to cure the shakes, but my one complaint with them was that they never stopped my head from racing. My heartbeat would slow down and my muscles would relax, but all the while, panicked thoughts continued to swirl, banging around inside a useless nervous system.

Still, there weren't many other options, so I got off the couch to head upstairs. But just then, my newfound green mechanism kicked in. It occurred to me there actually was another option: I could tough it out. I could breathe with my stomach, like my yoga instructors always said to do, drink some relaxing chamomile tea, and remind myself of all the things I still have, like my health, my family, my friends, and my job. Not only this, but I could kill two birds with one stone and count it as one of my changes—I could green my nervous breakdown.

I quickly logged onto my blog and typed up a short post, explaining that I'd "suddenly found myself at the buffet of life with a mountain of stuff on my plate, and unfortunately a lot of the good stuff was underneath a lot of pickled beets." A lame metaphor, but the point was that I was going to overcome my panic attack the natural way instead of with prescription meds.

Just as I was locating the Wikipedia entry on Lorazepam so I could embed the link into the URL on my post, Meghan phoned.

"How was your day?" she asked, cheerily.

"Uh, well . . .," I muttered, then broke down sobbing yet again. This was apparently going to require buckets of tea and a marathon yoga session.

Unlike Ian, however, Meghan is incredibly responsive when it comes to heartbreak, and she's as sympathetic as one can get. Within ten minutes she was at my door bearing hugs and a jar of homemade lentil soup, consoling me and offering to ghostwrite my blog post for the next day. I told her not to worry about it, but she went and typed up an emergency post, just in case, when she got back home, about enforcing regular quiet time and leaving the stereo, computer, and cell phone off. Meanwhile, I heated up the soup, ate it with my tea, and felt much better.

Still, it wasn't enough to stop me from throwing on my puffy vest, shuffling over to the 7-Eleven, purchasing a family-size bag of Crunchy Cheetos and a pack of cigarettes, coming home, opening a bottle of red wine, sitting back down, and making my way through the bag, the bottle, and half the pack before collapsing in my sweatpants and pink flannel pajama top sometime around 2 a.m., not caring one bit that I'd broken at least seven green rules.

NOVEMBER 15, DAY 260

Support charities without wearing wristbands or ribbons

Whether as passive-aggressive punishment for going on a six-week vacation or because he thought taking on one personal challenge and blogging about it wasn't enough, my editor assigned me a story on this new book called *A Complaint Free World*, for which I'd have to interview the author and take him up on his twenty-one-day challenge. This meant that I would now be making all of my green changes while trying my best not to complain, whine, gossip, or criticize anyone or anything, and of course I'd also have to document the experience on the *National Post*'s new arts and life blog, *The Ampersand*.

Two blogs. Two personal challenges. And no whining? This was highly unlikely.

The most objectionable aspect of the whole complaint-free thing, however, was the fact that it had its own rubber wristband, much like the yellow LiveStrong bracelets made popular by Lance Armstrong in his efforts to raise awareness about prostate cancer. Wearing an accessory to communicate the fact that cancer sucks and we should maybe do something about it is one thing, I thought, but a wristband to say nothing more than "I hate complaining . . . not that I'm complaining about it" is another. Although, to the man's credit, the purple bracelet could also be used as a tool of self-flagellation, much in the way one might snap a rubber band against one's wrist at every swear word or nail bite—this was actually suggested on the website. Either way, 5,405,203 of these bracelets had been shipped so far to eighty countries worldwide.

After complaining about this complaint-free ridiculousness in the days leading up to the challenge—I mean, honestly, how was I supposed to review a film without criticizing it?—I realized the author did make one good point: if something is causing stress, depression, or anger, one should attempt to change the situation instead of just griping about it.

This gave me an idea. I decided that, for today's green change, I'd refuse to wear any rubber wristbands, ribbons, brooches, or badges, no matter what the cause. This would save precious rubber plants, prevent artificial dyes from being used, and lower demand for shipping, manufacturing, and all the other carbon costs associated with such accessories.

However, I would make one exception—my new Habitat necklace, given to me by my mother.

As part of a change I'd made in July—my commitment to volunteer on a regular basis for environmental groups—I decided to look into a project called Women Build run by Habitat for Humanity, where women of varying ages come together and help construct houses for single mothers in need (all while wearing stylish pink hard hats, of course). My mom had heard about it through a neighbor, so she suggested that she, my sister, and I go to an information session tonight.

It was being held at Verity, an upscale women's club downtown. I arrived on my bike, and for the first hour or so didn't do much other than stand around with my helmet under my arm, drinking glass after glass of lukewarm Ontario Chardonnay—Ontario and Chardonnay being two descriptors I normally avoid at all costs in my oenophile pursuits but was having to accept into my vocabulary more and more these days thanks to the local alcohol pledge. Just as the third tray of shrimp tempura went by and my salivary glands began dripping with resentment toward my green rule about consuming ethical fish, our host finally emerged and escorted us into the meeting room for a presentation.

It was mostly straightforward and perfunctory, but near the end, a woman named Princess Water came up to speak, and single-handedly brought the tearjerker genre to a whole new level.

In the late eighties, she and her husband, Prince, escaped war-torn Liberia to come to Toronto, where he was soon diagnosed with a rare throat infection. He passed away in 2002, leaving her to raise eight children by herself. In the meantime, she learned that her brother back home in Africa had been killed and the family home burned to the ground.

After applying for subsidized housing, Princess waited seven years before being offered a run-down, cockroach-infested place in Scarborough, on the eastern outskirts of Toronto, right by one of the city's most violent, gun-plagued intersections. She asked the housing commission to do something about the cockroaches, so they eventually came to spray the place, but soon after this, her youngest kids began to develop asthma. The heat wasn't functioning, either, so the family was forced to struggle through the harsh winters by wearing layers of jackets, hats, and mittens inside the apartment. Often, as she commuted to and from work on a bus that crossed over a cement bridge and back each day, she wondered if she would eventually get off and jump.

A friend, after watching her suffer, told Princess about Habitat for Humanity. She wasn't convinced she'd be a worthy candidate, but filled out an application anyway. To her surprise, she was chosen, and now, after some five hundred hours of hammering, drilling, and plastering with the Habitat team—as well as other volunteer work with the organization and lessons in finance, insurance, mortgages, and various legal responsibilities—the single mother of eight has a home of her own.

To boot, it was one of the first Energy Star homes built in the community, which means it's 30 to 40 percent more efficient than most other single-family houses built to minimum Ontario building code standards.

When Princess finished speaking and stepped down from the podium, I looked over at my mother. She was wiping a tear from the corner of her eye. I looked back at my sister, who whispered, "They really know how to advertise."

I rolled my eyes and joined in the applause. On my way out, I signed up for the next Women Build project and left my e-mail address. I figured if I could construct a compost bin from scratch, surely a house couldn't be that hard—I even had my own staple gun, not to mention some connections in the lumber department at Home Depot. My sister signed up as well, while my mother did what so many women do when in a highly emotional state: spend money. Some of the organizers were selling silver and gold necklaces with little key pendants; half of the proceeds were going to Habitat. She bought one for Emma and me as an early Christmas present, and while I wasn't sure where the gold was mined and whether it was done so ethically, the charitable aspect of the necklace meant, in my opinion, that I wasn't technically breaking my rule about eco-friendly and/or fair-trade jewelry.

"So are you going to sign up to work on the construction site, too?" I asked my mother, after thanking her for the necklace.

"Oh, no," she said. "I'm better with a wallet than a hammer."

NOVEMBER 16, DAY 261

Reuse old computer discs as coasters

One of my best ideas for reusing something: turning old floppy discs into drink coasters. Instructions: place drink on disc. Done.

As much as I hate the term eco-chic, this comes pretty damn close to representing it.

I just wish I'd brought a few of these along to my interview with chef Jamie Oliver this afternoon, because he insisted on buying three drinks before we even began—one for himself, one for me, and another for the photographer.

His drink, a sidecar, came with a big sour gummy as a garnish. After taking a bite from the end, he offered to share the remains of it with me.

Jamie Oliver. Offering me part of his sour gummy. I knew it must have been full of food-grade petroleum, refined sugar, and everything that didn't fit into my current food restrictions, and I knew that I'd been breaking more than a few of my green rules of late, but if I turned down this opportunity to basically swap spit with Britain's hottest celebrity chef (at least after Nigella Lawson), I also knew that I'd regret it for the rest of my life.

"Thanks!" I exclaimed, and immediately shoved the entire thing into my mouth, at the same time realizing I probably should have asked my question first.

NOVEMBER 18, DAY 263

Drink straight from the bottle

With all the changes I've made so far, I'm finding that I'm surprisingly comfortable with much of the hippie lifestyle. Shopping at thrift stores? No problem. Eating stuff made with ingredients like amarynth flour and quinoa? Great. But the snob in me hasn't quite died yet, which is why I find some of these seemingly inconsequential changes so unbearably difficult.

Today's change is one of these. Drinking wine from the bottle instantly brings back memories of getting smashed on white zinfandel in the living rooms of whichever parents were away during high school. It reminds me of a time when I would drink for the sole purpose of getting drunk, when coffee tables would serve as dance floors, when a glass just seemed like an unnecessary middleman and I could be inebriated without knowing how to spell it.

I'm making this green change now to avoid using a drinking vessel and thus prevent using more water and dish detergent to clean it afterward, but there's just something wrong about having beautiful crystal stemware sitting patiently in the cupboard, waiting to get acquainted with some vintage Baco Noir, while I proceed, instead, to swig the stuff straight from the bottle, leaving mucky traces of dinner around the rim.

My way to temporarily dodge this uncouth restraint is to go out for dinner. The majority of my minor food-related pledges are made with the caveat that, if I'm out at a restaurant, all bets are off. I will always search the drinks menu for local wine, for instance, but if the establishment doesn't offer any, this doesn't mean I'm not having wine, it just means I'll opt for the Californian over the Chilean variety. And if the meat isn't from a sustainable farm, I'll go for a pasta dish; however, if it includes cheese that isn't organic, I eat it nonetheless. Of course I try to request that my friends and I frequent vegan or vegetarian restaurants and attempt to cook at home as much as possible, but this doesn't always happen. Tonight, for example, after seeing a movie with Ian and another close friend, Dimitris, we were desperate for dinner and the boys were craving sushi. There was a place right across the street, so I went along, hoping for some vegetarian options.

The restaurant had a sweet potato and avocado roll, so I ordered that, knowing full well that avocados probably weren't being grown in the Greater Toronto Area in November, and topped it off with some even less local beer—Asahi, from Japan.

I let out a sigh of guilt and disappointment, apologizing to Ian and Dimitris for officially breaking two of my rules.

"Don't worry, we won't tell," said Dimitris. This is, sadly, now a common refrain among my friends, said to reassure me if I'm ever in a situation where I have to break a pledge.

Ian then asked me whether I felt guiltier about being dishonest to my readers or about the actual carbon cost associated with consuming avocado and imported beer.

After thinking for a bit, I answered that it was more to do with the former: for all my thousands of readers, friends, and family knew, I was being a perfect greenie, keeping up with every single change so I could prove to the world that this lifestyle was, in fact, completely doable. And while I'd come to truly loathe certain ungreen things such as plastic bags, disposable coffee cups, and overpackaged individual portions of food, I really hadn't come to feel any animosity toward tropical fruits or hot baths.

People have already started asking me what I'm going to do when this year is up—what changes I'm going to keep and which ones I'm going to scrap—and although I can't really answer that until it happens, part of me knows that I'm not about to turn away good food just because it arrived on my plate via airplane.

"So do you feel like a hypocrite whenever you screw up?" asked Ian with characteristic bluntness, making Dimitris chuckle through his sashimi.

"Uhhh," I said, caught off-guard. I wasn't all that sure. Did I feel like a hypocrite? Was I one?

NOVEMBER 20, DAY 265

Knit my own scarves and mittens

Knit 1, purl 1, knit 1, purl 1 . . . This is the pattern I'm following for my new winter scarf, made with unbleached, unpackaged wool from a yarn store down the street. It's as local and sustainable as winter accessories get, unless perhaps I collected my cat's shedded fur, combed out the dander, spun it on a loom, and went from there, which is an activity I think I'll save for my crazy cat-lady years.

The great thing about knitting is that it's therapeutic and simple to do, once you get going. The key, my mother taught me, is to keep your tension consistent, making sure not to wind the thread tightly for one stitch and loose the next.

This won't be a problem, however, at least not for today, because every single stitch I make will be wrapped so tightly around the needle you'd think I was trying to strangle the damn thing. See, I've got some pent-up anxiety, having suddenly realized that my much-anticipated one-hundred-day countdown to the end of this green challenge can't actually start for another twenty-four hours because it is, just my luck, a freakin' leap year.

Unless my calendar, my weekly planner, my computer, and the millions of Google hits that came up were wrong, I apparently have 366 changes to make, not 365. Yes, it's just one more, out of hundreds, and really it shouldn't bother me that much. But it does. It's like Mother Nature and the Gregorian calendar are both ganging up against me. And after all I've done for them! Well, one of them, at least. Maybe as my next green change, I should switch over to the lunar calendar.

NOVEMBER 24, DAY 269

Buy and refurbish used furniture

I'm standing with my parents in a suburban warehouse, gazing at a never-ending row of ugly curtains. They're made of translucent gauze and look like super-sized versions of the skin you peel off after a week of suntanning.

Oddly, I feel like I've seen them before.

In fact, it isn't so odd; I probably have seen them, or something like them, whenever I've been traveling and staying at three-star hotels—this is where they all come from, hotels. Specifically, they're the thin curtains that are hung between the window and the usually darker, heavier drapery.

But the warehouse is full of more than just this. It has an entire wing devoted to outdated teak and oak desks, a back room with bed frames and mattresses, and a main floor, which is littered with poofy and faded floral-motif couches, armchairs, and ottomans, the sort that are offensive in their sheer inoffensiveness.

My parents drove me here—I give my mother full credit for finding the place—as I'm planning to buy only used furniture for the new house, and this dusty hotel stuff is perfect for recovering. Each chair is only $40, the curtains were around $12. Nothing much was selling for over $50. All it would take was a few yards of fabric, a decent upholsterer, and maybe a week or two of patience, and I'd have myself some brand-new—and yet not really new at all—living room chairs.

Later on, when I got home, I flipped open the Yellow Pages to look for an upholsterer. I had no idea which company to call, so I did what most people do in this situation and phoned the first number on the list: Akram Upholstery Ltd.

Akram, or Mohammad, as his first name turned out to be, was this funny man in his late fifties who'd been in the upholstery business for decades and worked out of a dilapidated clapboard house on dreary Dufferin Street.

The moment he picked up the phone, I could tell he was, as some people are fond of saying, "a character."

"Van-eh-ssa, Van-eh-ssa . . .," he said, about thirty seconds into our conversation, after I had described the chairs and asked for a quote. "You know I will give you good price. You know why I give you good price?"

No, I said, I didn't know.

"Because, Van-eh-ssa . . . because you are beautiful."

Oh lord. This could get embarrassing.

"How do you know I'm beautiful?" I said. "You haven't even seen me before."

"I can tell, I can tell—by the sound of your voice."

Mohammad then asked me to describe the chairs again and made a suggestion for the amount of fabric I should get. Then he asked when I needed them by and I replied that it wasn't a rush, they could wait until after Christmas.

"Oh, Van-eh-ssa, this is wonderful, this is wonderful," he exclaimed, "because you know, I have some other projects right now and it gets very busy for me at this time. But for you—for you I will work extra long hours."

"No, no, honestly," I said, "don't worry about it. Really, please don't work overtime for my sake. I have a couch. It's a great couch. I don't even need the chairs until February anyway, so take all the time you want."

"Oh, you are too kind, too kind!" Mohammad said. "These chairs, these chairs are going to look so beautiful. Not as beautiful as you, of course, but almost as beautiful."

I was starting to feel some pressure, now, to actually look beautiful—far more beautiful, even, than whatever fabric I got for these chairs. This of course meant I'd either have to buy some really dull fabric samples or remember to reapply my lip gloss and eyeliner before heading over to Mohammad's.

Just as I was thanking him and trying to wrap things up, he asked me why I'd decided to reupholster old chairs rather than get new ones. Without wanting to explain my whole challenge—a spiel I had, by now, condensed into less than thirty seconds but which inevitably required twenty minutes of follow-up Q and A—I simply said that I wanted to be environmentally friendly.

Little did I know, Mohammad fancied himself a bit of an environmentalist. He'd even printed this on all his business cards, which he told me said, "Our philosophy is recycle your furniture and save trees" in green, all caps, with a picture of a tree on one side and a recliner on the other. He then gave me a thorough run-down on all the ways in which his business practices were eco-friendly.

It was sweet, really. But I was also getting charged by the minute for this call and had to interrupt him. I tried to think up some environmental reason for why I needed to hang up; in the end, however, I resorted to the fail-safe.

"Oh no, my battery is about to die, Mohammad!" I said. "But you can tell me all about this when I see you next week to drop off the chairsokaybye!"

NOVEMBER 29, DAY 274

Use coconut oil to polish my shoes

For the sake of both curiosity and irony, I'm really tempted to lick my own boots right now to see if they taste like piña colada. Even if they don't, I'm sure they'll taste better than turpentine, ethylene glycol, lanolin, gum Arabic, wax, and naptha—a flammable mixture of hydrocarbons from petroleum or coal tar—ingredients that can all be found in commercial shoe polishes.

Definitely not as yummy.

december

1	No more tape
2	Don't shave my legs
3	Avoid products that contain corn from GMO/monocrop sources
4	No dish detergent unless there is an oily residue
5	Buy beans dry, in bulk
6	Have an "inside day" every couple of weeks
7	Add a green tip to my e-mail signature
8	Drink only fair-trade tea
9	No more individually packaged, single-serving foods
10	No more downhill skiing
11	Take only cabs fueled by natural gas
12	Eat only organic honey
13	Hire couriers who use bicycles and public transit
14	Drip-dry dishes in dishwasher rack above houseplants
15	Hand-whip my whipping cream
16	Buy organic, unbleached cotton towels
17	Use exact change whenever possible
18	Use a broom and dustpan instead of a vacuum
19	Pack light
20	No more paper towels
21	Use scrap paper for bookmarks
22	No more power tools
23	Wear only natural lipstick
24	Stay organized to avoid losing (and repurchasing) things
25	Use bathroom before boarding a plane
26	Use a crank-up radio
27	No more highlighters
28	Make my own cosmetics and beauty products
29	Use only old boxes and storage containers
30	Buy used appliances and kitchenware
31	Use cold water for washing hands, face, and dishes

DECEMBER 2, DAY 277

Don't shave my legs

I AM NOW, officially, a hippie. Is there a membership card for this? A hemp necklace, at the very least?

DECEMBER 3, DAY 278

Avoid products that contain corn from GMO/monocrop sources

I continue to be blown away by corn. It's everywhere—not just in our salad bars and in popped form at the movies but in our cans of baked beans, our hamburger meat, and almost every dessert, not to mention in biodegradable takeout containers, supposedly eco-friendly ethanol, and even in our hair. After watching this indie documentary *King Corn,* which was partly inspired by the first part of Michael Pollan's book *The Omnivore's Dilemma,* I bolted over to my pantry and began madly poring over every ingredient list on every prepared food item I had. Anything that included corn syrup, cornstarch, glucose, or fructose meant it had corn—and not the healthy kind you see in the produce aisles of the supermarket every fall but the genetically modified kind that produces multiple ears per stalk, the kernels of which are almost inedible until they get processed and refined. I mean, I know it's good to be all-ears, but not when it comes to food.

Along with soy and wheat, corn is also one of the most prevalent and environmentally taxing monocrops in North America, so I'm going to try to avoid it in all its forms starting today. I might even put a limit on my corny puns.

DECEMBER 7, DAY 282

Add a green tip to my e-mail signature

For a while now, Meghan has been indulging in a bit of discreet self-promotion by sending all her e-mails with a cutesy signature tacked on at the bottom, below her name and address. It says: "Pass me the brown rice please," and next to this is the link to her blog.

Then today, I noticed that the editors at Treehugger.com had written a post about incorporating green tips into e-mail signatures and I thought this might make for a worthwhile change. After all, I had three e-mail accounts—work, personal, and blog—and probably sent at least twenty e-mails a day, so at least a few people would notice the message at the bottom.

Treehugger's suggestion, for people who cared enough about the environment to do this but didn't have the faintest idea what to write, was, "Eco-Tip: Printing e-mails is usually a waste."

Eco-Tip: Printing e-mails is usually a waste?

Way to bring on the rhetoric, Treehugger. For as long as I live, I will never print out another e-mail again, thanks to your sage and originally worded advice.

In fact, I can't even remember the last time I actually needed to print out an e-mail. Does anybody still do that? Anybody besides lawyers and septuagenarians, that is?

Of all the tips to write, that sounded pretty lame. Besides, I'm a writer, so if I don't come up with something caustic and witty, I could get exposed and lose my job . . . or at the very least get the piss taken out of me by my colleagues.

I gave it some thought, but it was more difficult than I'd predicted, mostly because it had to remain professional. Also, it was important to choose a tip that people might actually remember and make a point of doing, which meant I couldn't exactly write a long-winded message about the importance of harvesting rainwater. I didn't want it to sound preachy, but it also couldn't sound trite, and it had to be tongue-in-cheek but still genuine. After much huffing, puffing, and hangnail-picking, I finally typed the following:

Vanessa Farquharson
Arts & Life Reporter
National Post
300-1450 Don Mills Rd
Toronto, ON M5B 3R5
t: (416) 383-XXXX
e: greenasathistle@gmail.com

** Green Tip of the Day: Get your morning coffee in a reusable thermos instead of a disposable cup. And yes, that includes the Grande Extra-Hot Non-Fat Latte With Foam getting cold on your desk right now.

————

Ugh. Disappointing.

But I did feel strongly about using proper thermoses, and this was a change that nearly everyone could make without much effort. I thought the run-on adjective was a cute but sympathetic dig at the popularity of Starbucks and showed that I wasn't necessarily going to ream anyone out for not drinking fair-trade coffee, and overall it was still pretty short and emphatic. At the very least, it was a whole lot better than trying to enlighten people with the fact that printing e-mails can often be a waste of paper.

DECEMBER 12, DAY 287

Eat only organic honey

A couple of weeks ago, I received a frantic message on my answering machine from Mark, requesting my zip code. I had called back to tell him what it was, then he hung up on me, which was somewhat confusing, but in the end I just forgot about it.

We'd been keeping in touch semiregularly after breaking it off last month—I remain friends with most of my exes, so this wasn't too unusual—but it was nonetheless an odd phone call. Also, I was pretty sure that Mark was in Hawaii at this point; in any case, the area code on my display screen certainly didn't belong to Portland. For one panicked moment, I thought maybe he was stranded on a remote island in the Pacific or being held hostage by the Polynesian mafia, but then why would he need my zip code?

However, it all became clear when, in today's mail, I received a package with Mark's name and a mailing address in Hawaii scrawled in the top left-hand corner.

It was something small, yet heavy, and it had been torn open already by a Canada Customs employee (a yellow sticker explained this). I reached in, wondering what on earth it was.

Turned out, it was honey.

I was still confused. Why honey?

"It's Volcano Island honey—it's the best," said Mark, when I called to thank him and subsequently question him about this post-breakup gift. I mean, I like honey and all, but was this supposed to mean something beyond the fact that he saw a jar of honey while he was in Hawaii and thought of me? Was I supposed to interpret this as a sign of his enduring sweetness and ask to see him again?

"Just try it," he said. "You'll see."

So I dipped my pinky finger in the jar, raised a blob to my lips, and prepared for the familiar taste of honey.

It was good. Actually, it was great. No, hold on—let me rephrase this: it was slap-me-across-the-face-please-because-I-think-my-taste-buds-just-took-ecstasy-and-had-an-orgy-in-my-mouth great. Every food cliché in the book couldn't even begin to describe the absolutely mind-blowing gastronomic pleasure this was giving me. God bless bees, and their need to barf.

"No wonder customs wanted to search this," I said, suddenly feeling in need of a cigarette. "Are you sure this is legal?"

He insisted it was, although it was raw and unpasteurized, which gave it a unique opalescent sheen and soft texture. The honey also came from an organic farm where the bees pollinated only a single species of flower.

After hanging up and thanking Mark again, more profusely this time, I went online to see if this honey farm had a website. It did. They had links to articles by the likes of *National Geographic*, which had in fact referred to it as "some of the best honey in the world," and the owners wrote extensively about their labor-intensive collection and extraction process.

I was reminded of all the news earlier this year about Colony Collapse Disorder and thought I recalled something about how the organic bee population was unaffected by it. I wanted to know more.

And yet, the prospect of researching honey was far less appealing than the process of eating it. It was time to delegate.

At the beginning of the month, my agent had suggested I get myself an intern to help keep me up to date on all the enviro-news and offer ideas for green changes. I posted an ad at the local university and a girl named Eva responded immediately. Her first assignment, I decided, would be to get the facts on honey.

Within the hour, she'd e-mailed back with surprisingly fascinating information about bees.

Colony Collapse Disorder, the name given to the sudden disappearance of 30 to 40 percent of the bee population in the United States and certain parts of Quebec this year, didn't translate into only higher honey prices but higher food prices, the reason being that a third of all food comes from crops pollinated by bees. But while commercial hives suffered major losses, organic beekeepers never reported any trouble.

When honey is processed and pasteurized commercially, it is finely filtered and heated, which removes pollen particles and even vitamins. Raw honey, however, retains the pollen, vitamins, digestive enzymes, and antioxidants.

When it came to organic versus commercial honey, research seemed to show that organic was better for the consumer's health, for the health of the bees, and for preserving pesticide-free plants.

"Also," Eva wrote, "because certifying a hive as organic is costly, the beekeepers don't usually exterminate the bees at the end of the season, which is a common practice in conventional honey farms.

"By supporting these beekeepers, you're helping maintain a population of healthy bees, without which many crops wouldn't be able to grow," she said, adding: "We really need bees!"

We *do* really need bees, I thought, suddenly feeling vindicated in my use of vegan waxed floss.

Just as I was about to hit "Publish" on what was surely the first thoroughly researched and comprehensive post on Green as a Thistle, I got an e-mail from my mom.

I'd told her about the gift Mark had sent and how it might turn me into a honey snob—or, as I'd prefer to be called, a melissopalynologist (person who studies honey).

She e-mailed me back, then, responding to my newfound obsession in the best way imaginable—with scientific backing.

"Thought you might be interested in this," she wrote, and underneath was a link to an article in an online medical journal about a study on the various uses of honey as an alternative topical treatment for wounds. I'd heard about it being used for sore throats before, thanks to its similar consistency to cough syrup—but wounds?

Reading it, I discovered the reason for this was because honey not only prevents germs from penetrating the surface of a cut, it also has antibacterial properties because of its low water content, as well as this thing called "the hydrogen peroxide effect," and high acidity.

Scanning through the Wikipedia entry on honey, I also read three random but fascinating factoids: 1) It's thought that people in the Roman Empire were allowed to pay their taxes with gold or honey; 2) the Ancient Egyptians used honey for embalming the dead; and 3) honey is the primary ingredient in mead.

Mead. Now there's a drink. Whatever happened to mead, anyway? I'll bet there's totally a market for local, organic mead.

DECEMBER 14, DAY 289

Drip-dry dishes in dishwasher rack above houseplants

"You should really look into a patent for that," said my Dad.

I had just found a use for my empty, unplugged dishwasher. Handwashing my dishes meant letting them drip-dry on a tea towel, which in turn meant the towel got mildewy after a while and had to be cleaned. I needed a drying rack but didn't want to buy any new plastic, so in the end I opened the dishwasher door, pulled out the top rack, and let the dishes dry there. As they sat, dripping onto the open door below, I got the idea to stick a few houseplants underneath so they could catch the water.

This setup meant that a little more graywater got used while one less tea towel went in the laundry pile. I named my invention "the dishwatering can."

My dad was right, there was something here—something that with a bit of research and development, consumer testing, and patent applications could really turn a profit.

Too bad I'm far too lazy to get on that.

DECEMBER 17, DAY 292

Use exact change whenever possible

The reasoning behind this one is that the more coins kept in circulation, the fewer coins need to be churned out by the Royal Canadian Mint; and fewer coins means less demand for mining copper, nickel, and other metals.

This also means you do not want to stand behind me in line at the cash register anytime soon.

DECEMBER 18, DAY 293

Use a broom and dustpan instead of a vacuum

Giving up my vacuum. My most dreaded change. It was the only thing left on The List today, and I couldn't come up with anything else to do in its place, but I'm already distraught just thinking about all the lint, crumbs, and stray hairs that will accumulate around my apartment, not to mention all the dust mites I can't even see.

I think I need a shower. A dark, lukewarm shower.

DECEMBER 19, DAY 294

Pack light

An ex-boyfriend of mine believed that packing was an art. A perfectly executed packing job—whether it was a bunch of clothes tossed in an overnight bag or an entire apartment's worth of boxes and furniture getting crammed into a U-Haul—was something one could learn with time and patience, but it was also an inherited skill. (He insisted that good packing ability happened to run through his father's side of the family.)

It sounds silly, but I truly did admire how he could make the most awkward and bulky of items fit into the smallest of suitcases, as though they were specifically designed for this purpose. I'm sure if I'd asked him to squish my entire desk into an already full duffel bag, he'd somehow make it work.

I picked up a few tips from him over the years, and I'm hoping to put these into practice when I get ready for a minitrip to New York between Christmas and New Year's. Ian and I are going together and will be gone for four days, flying—yes, flying—there and back. My plan is to smoosh everything into a carry-on bag because the lighter the luggage, the lighter my carbon footprint (less weight on the plane means less fuel is required, which is a minuscule difference, of course, but I'll take what I can get).

As I started to think about exactly what I was going to pack and how it might all fit in one tote bag, my phone rang. The caller ID said it was Jacob, phoning from his dad's house in Toronto. He wasn't supposed to be back in the city for another two days, which I knew because on my calendar I'd written "Jacob back," with a little smiley face next to it, on Friday.

I flipped my phone open.

"Oh my god, you are *not* back already!" I exclaimed.

"I *am* back already!" he said, explaining that his office had let him leave ahead of time and he was able to find an earlier flight. This is the other thing about Jacob: he has a mild obsession with flight schedules. And by mild obsession, I mean that he's memorized the times of nearly every flight from every airport with every airline and can often provide prices, accurate to within $50, on any given day. So, in fact, it's actually not so unusual for him to appear in Toronto two days early thanks to some obscure 3 a.m. flight on Brussels Air from Tel Aviv with a stopover in the Canary Islands.

"I'm unpacking," he said.

"Really? I was just thinking about packing," I said.

"You were? For where? You're not going to New York yet, right?"

Jacob was originally going to come along with us on the trip but his dad and sister had made plans to take him up north, and considering he rarely gets to see them, he backed out.

"No, no, not until after Christmas," I said. "Hey, I'm going out for dinner tonight with my parents and I think Dimitris is coming along. You want to join us?"

"Yeah, sure—seven o'clock at Browne's, right?"

"You got it."

When dinner finally rolled around, Dimitris and I arrived first, and were soon joined by my parents and Emma. After about ten minutes, I was really hoping Jacob would be as punctual as he is knowledgeable and walk in the door because my mother was somehow already tipsy and had for no explicable reason begun asking Dimitris when he last had his heart broken. I'd interrupted her twice now to try to steer the conversation in another direction but she was having none of it. Dimitris, thankfully, was taking this all in stride and attempting to give my mom a vague yet satisfactory reply about the meaning of heartbreak in general, but it was nonetheless getting awkward.

Finally, in walked Jacob. It was funny—that night three months ago, when Emma and I first arrived in Ramallah, I'd felt incredibly desperate to embrace the boy and attributed it to our being halfway across the world in a conflict-plagued region at two in the morning. Now I suddenly had the same feeling. But was I really that desperate for him to save Dimitris from my mother's persistent interrogation? Maybe it was just that I hadn't seen him in a while and the whole home-for-the-holidays thing was getting to me. Either way, I got up to give him the best welcome-back hug that I could.

Dimitris followed suit, then my family all did the same. Jacob eventually sat down, picked up a piece of bread, and started a fresh line of conversation with my dad about the trials and tribulations of starting a business in the Middle East.

I sat down, too, and listened to everyone talk, but I still felt all restless and fidgety, as though there were something I hadn't quite finished or wanted to say.

I ordered a glass of red wine, thinking it would at least help me relax, but by the time it had been uncorked, poured, and plonked down in front of me, I realized that what I really wanted wasn't wine at all. What I wanted was another hug.

DECEMBER 23, DAY 298

Wear only natural lipstick

There's a lot of talk these days about greenwashing—the tendency for companies to make claims of being eco-friendly in some way when, actually, they're just selling a slightly less carcinogenic product, or have repackaged something in a more concentrated format so the consumer needs less of it each time.

But on the flip side, there are some companies and products that are so ridiculously green it's almost overwhelming. Take this lipstick that I just bought today: it's a brand called PeaceKeeper Cause-metics (cute, huh?), which makes lipstick that's mineral-based and 100 percent all natural. The ingredients are detailed on their website, and the company is ranked as one of the safest cosmetics manufacturers in America. All the profits from the sale of these lipsticks are donated to women's health advocacy and human rights organizations. Their packaging is recyclable, and, in another year or two, they'll probably have figured out how to make the actual lipstick tube biodegradable, as well. (Cargo has done this already, in fact, with their Plant Love line of lipsticks, which come in tubes made from that eco-friendly and yet not so eco-friendly ingredient, corn.)

It's at once commendable and disturbing how much ethical consideration can go into a single product like this—perhaps the folks at PeaceKeeper should think outside the lipstick tube and look into volunteering their services to the oil industry.

DECEMBER 25, DAY 300

Use bathroom before boarding a plane

Airplane toilets have nothing to do with Christmas, of course, but I'm in the mood for non sequiturs today and will be encountering mile-high toilets in less than twenty-four hours, when Ian and I fly to New York.

The last time I was in Manhattan was about ten years ago with my parents, and I can't remember much about the trip other than having to buy earplugs to sleep at night and being utterly perplexed by the popularity of the Jekyll and Hyde–themed restaurant. This time, I had a list of things I wanted to accomplish: 1) eat a Magnolia cupcake; 2) spend too much money at Anthropologie; 3) spot a celebrity; 4) form an opinion on Brooklyn; and 5) meet No Impact Man. Perhaps, if time permitted, we could visit MoMA and see Times Square, but cupcakes and celebrities came first.

None of these things on the to do list were especially green, of course, other than maybe meeting up with Colin, my fellow eco-blogger. In fact, a fair amount of what I wanted to do in New York required breaking numerous green rules (the cupcake frosting, for instance, would probably have corn syrup in it, and the clothes at Anthropologie surely wouldn't be sweatshop-free). But as the tourism industry likes to say, the point of a vacation is to get away from it all—and in my case, "it all" includes the green stuff. I know that if I save up the money and spend the time to go to New York, then actively restrain myself and try desperately to *not* do all these things I want to do, it's only going to lead to stress and resentment.

In fact, I'm realizing more and more these days that making occasional nongreen indulgences is an integral part of being a tolerable environmentalist.

If I spend about 90 percent of my time living in accordance with my established green values—eschewing processed and imported foods, going to farmers' markets, choosing organic and local produce, and so on—and use the other 10 percent of my time to make conscious but not so environmentally friendly choices, such as eating cupcakes made with refined sugar, food coloring, and corn syrup, there will ultimately be a feasible balance of virtue and vice. Some environmentalists who predict an apocalyptic global warming disaster in the near future will argue we don't have time for any eco-vices, that we all need to implement massive changes and stick to every single one of them if we have any hope of surviving, but personally, I know I can't survive without a few sins here and there.

Don't get me wrong: I realize that if I went back to my old lifestyle—the one in which I had minimal regard for my purchasing habits beyond their monetary cost, and even less regard for things like the carbon footprint of a cupcake—I'd be left stranded in bitterness and cynicism. But trying to be Gaia's prodigal daughter, striving for absolute eco-perfection, would only make me neurotic; I'd obsess about every detail, lose sleep over whether CFLs or LEDs were the way of the future, suffer debilitating guilt every time I flushed the toilet, and subsequently alienate most of my friends and family. The thing is, there's no hope of convincing people to change their lives for the greener as long as they can look at an environmentalist and see nothing but a one-dimensional tree-hugger ready to pounce on every strike against Mother Nature.

This need, then, for both vice and virtue, skepticism and earnestness, is what I keep mulling over, especially when I'm forced to confront and rationalize my hypocrisy during this challenge. When Ian called me out at the sushi restaurant last month for eating imported avocado and beer at dinner, I spluttered a retort about how it was the greenest choice on an entirely un-green menu. When he brought up the flight we'd be taking to New York, I countered that it wasn't technically hypocritical because I wasn't going around telling people not to fly, I was just telling them to bring their own headphones on the plane. But regardless of the semantics involved, the fact remains, it exists—hypocrisy is everywhere, it's unavoidable, and the more I think about it, the more I believe that, in small doses, it's actually quite healthy.

When I finally sink my teeth into that Magnolia cupcake, there's no doubt that I'm going to taste more than just sugar and frosting—I'm going to taste monocrops and guilt. But in a bizarre kind of way, that can sometimes taste just as sweet. As much as it's essential that I move onward and upward in this green challenge, I also have to let myself fall down every now and then. It's a philosophy that's more *Que sera sera* than *Carpe diem*, but in the long run, I'd argue it's far more sustainable.

Which brings me back to airplane toilets.

It's inevitable that people are going to keep flying—and I'm one of them—so I've decided that another small but no less relevant change I'm going to make to my air travel (along with requesting a vegetarian meal, packing light, and bringing my own set of headphones) is to pee before I board the plane.

An article on Treehugger recently discussed the fact that while these sky-high toilets are fairly water-efficient, relying on mostly suction power, the energy used in one flush still amounts to over fourteen pounds of carbon dioxide, enough to power an average-size car for six miles. Plus, technically, if everyone on the plane has a full bladder, it weighs more, which in turn requires more fuel to propel the aircraft forward.

Bearing this in mind, I'll be more than happy to abide by the rules and not carry any liquid whatsoever onto the plane.

DECEMBER 26, DAY 301

Use a crank-up radio

"Burn out the day! Burn out the night! I can't s—

"... to put up a fight. I'm livin' for—

I was having so much fun with my crank radio, one of the green gifts my mom bought me for Christmas. There was still some time before I had to head to the airport, so I decided to keep hanging out at my parents' house. The station Q107 Classic Rock was playing one of my favorite retro songs, Blue Öyster Cult's "I'm Burnin' For You," and I hadn't yet gotten around to charging the battery, which meant the music would come out only while I was turning the handle. This made me feel as though all of Toronto's airwaves were now entirely in my control.

"God, this is like when you buy a kid a toy without thinking about all the noise it'll make," said my mother. "I can't believe I still haven't learned that lesson."

The radio, along with all my other gifts, had been wrapped in newspaper, in recognition of my rules about no new wrapping paper. My parents had gone so far as to make sure every single present I received conformed to my green standards—old Pyrex mixing bowls, an antique coffee tin and ice cream scooper, bamboo socks, mineral-based eye shadow and natural blush, beeswax candles, fair-trade hot chocolate—and had also bought a free-range turkey, organic root vegetables, local greens, and a homemade apple crumble for our Christmas lunch. The only rule I broke that day was the one I most often break, the one about drinking only local alcohol—I very willingly helped my family make their way through a bottle of Veuve Clicquot pink champagne, and to tell the truth, I have very few regrets about it.

Of course, receiving gifts is always easy. The hard part for me this year was buying gifts for my family that were green in some way but still represented something I thought they'd want. I couldn't exactly give my father a Tetra-Pak of Ontario wine, for example, nor my mother a hemp T-shirt, or my sister a reusable coffee thermos. They wouldn't want these things, and it would only make me look thoughtless.

In the end, though, I think I did pretty well: for my mother, I bought a South American family a working beehive and a beekeeping kit through the charity Heifer.org, as well as some used novels; for my sister, I got some natural cosmetics and a few pairs of retro but sexy underwear from American Apparel; and for my dad, who as usual was the hardest to figure out, I finally decided on a rather bizarre but without question unforgettable and experiential present—a butchering class. It would be held in late January at the Healthy Butcher, a gourmet store that sources all its meat from local, sustainable farms with grass-fed cows and free-range chickens; the carving and anatomy class would give us both a chance to see where, exactly, our meat comes from.

Not only was the green gift-giving a complete success, but when I went online to check my bank statement and prepared for the usual punch-to-the-gut shock of how much I'd spent this month, I was surprised to find that I was still very much in the black. The environmental restrictions on presents meant I had to stop and think before opening my wallet, so I didn't make any impulse purchases. But on top of this, as I scanned back through old statements, I realized my account had actually been increasing in funds for months now. A lot of my readers asked how I could afford pricey organic food, bamboo dresses, and designer silk-screened tote bags. The answer: no car.

Ever since selling it in June, the number of monthly withdrawals from my checking account had gone down by almost half. Even if I'd given away my Bugaboo for free, I would still be saving hundreds of dollars on all the gas, insurance, and other expenses I no longer had. The occasional splurge on a bamboo dress was nothing compared to the cost of repairing a broken indicator light or paying $15 every time I parked underneath the Manulife Centre for a couple of hours to see a movie. As well, there was the fact that whatever I bought these days would have to fit in my bike basket or be carried all the way home on the subway and streetcar, so I was far more aware of the sheer quantity of stuff I purchased.

In essence, it appeared that in the process of reducing my footprint, I had fattened my wallet.

DECEMBER 27, DAY 302

No more highlighters

This sounds like yet another superfluous change: no highlighters. But anyone who knows me knows that I like to be organized, like to compartmentalize, like to prioritize. This isn't necessarily a good thing; it often means I feel some discomfort when my social circles have to collide at a party, or when someone puts a book back on my shelf and doesn't stick to my alphabetical ordering system. I tend to get especially fastidious about books, too—at home, not only is my collection alphabetized, it's organized by genre, and all the spines are perfectly flush with the edge of the shelf. I also have to own every book I read, so I can have it at hand for future reference (which means no libraries, much to the chagrin of my hippie readers). When I interviewed John Irving and got him to sign my copy of *A Prayer for Owen Meany* and later snapped up signed copies of works by Salman Rushdie and Timothy Findley, I had to restrain myself from getting obsessed with autographed books; similarly, when I found a beautiful 1886 hardcover of *Tom Jones* for a decent price on eBay, as well as one of the first North American editions of *Lolita,* I had to make sure I didn't head down the slippery slope of antique book collecting.

In college, while making my way through one reading list after another, I employed the following study technique for each book: highlight the passages that seem important; put sticky notes with further analysis on the pages discussed in class; go back and underline the even more important phrases within the highlighted chunks that I might need to know come exam time. I wasn't so anal-retentive that I felt the need to color-code my highlighting, but I did like to switch it up between a classic yellow for textbooks and more vibrant greens, pinks, and oranges for novels. Never blue, though.

The point is, I like highlighters. Even though I'm not in school anymore, I still feel compelled to use them. Much in the way some people need to have a drink in one hand and a cigarette in the other, I like to have a book in one hand and a highlighter in the other. But as of today, this changes. I figure, if I read something and want to remember or return to it, I can just fold down the corner of the page and maybe underline a few sentences with my biodegradable pen.

I'm guessing I'll be doing a lot of reading in the coming months, too, because Ian and I have officially arrived in New York, and while he went off to check out a music store, I went over to the Strand on Broadway and am now feeling the intoxicating weight of eighteen miles' worth of literature—used literature, at that! I can get a massive hardcover of *Anna Karenina* for just $12 (not that I need another one), *The Brothers Karamazov* for $5, *Middlemarch* for $8. It's every bookworm-slash-environmentalist's dream.

I'd have to thank No Impact Man. He suggested I visit this bookstore after we met up for coffee around the corner. Obviously, there was no way I was going all the way to New York without confronting my competi—. . . I mean, meeting my fellow green blogger. Colin, at this point, had just finished his own challenge and was in good spirits. Some producers in Hollywood had bought the film rights to his upcoming book. He was busy doing television interviews and had wrapped up a documentary that was being filmed about his project. Yes, here was the difference between Colin and me: this green business of his was just that—a business; it was his full-time profession. I, on the other hand, was still trying to make money writing about whether or not Patrick Dempsey's latest romantic comedy was as horrible as it looked from the previews, all the while keeping over three hundred ways to be environmentally friendly in my head, and in practice. I'm not going to lie: there's a bit of jealousy on my part.

At the same time, though, I'm truly in awe of the extremes to which he went, not just individually but with his wife and daughter. And I was excited to meet him in person, too, if only to put a face to the No Impact name.

We met at his favorite haunt in Greenwich Village, the Grey Dog, which was cute and cozy but not especially green in any way—the coffee wasn't advertised as fair-trade, the dairy wasn't organic, and Colin said he had to spend some time convincing them to get rid of the plastic cups they handed out for the water cooler at the back. Still, there'd surely be something I could eat.

I knew vaguely what Colin looked like from a couple of photos on his website, but when I first saw him at the café, he seemed smaller in stature than I'd imagined, as well as more docile and soporific, with a gentle voice to match. He carried a tote bag he planned to use at the Union Square farmers' market afterward, and he walked—fittingly—with a light step.

"So you must be right in the thick of it now, huh?" he asked me, realizing I had only two months left to go in my green year.

"Yeah, totally," I said, hastily ordering a bagel with cream cheese, then suddenly realizing I'd just broken my rule about organic dairy. For a split second, I had a paranoid premonition of the headline on Colin's blog the next day: *The Truth About Thistle—'Green' Blogger Recklessly Eats Philly Cream Cheese.*

"Um," I tried awkwardly to explain, "I kind of bend the rules a bit when I'm on vacation."

He didn't seem to care.

As we sat down by the window, I brought out my handkerchief to use as a napkin and I remarked on his normal appearance and the surprising lack of astringent compost fumes trailing him. He admitted to buying clothes for the first time in a while and also tossing his compost bin because of a persistent fruit fly problem, then complimented me in return for my similarly average attire and lack of body odor (such are the bizarre forms of praise that occur among embedded environmentalists).

The ensuing conversation revolved mostly around what we'd

been learning throughout our green years, whether it had to do with the time-consuming nature of the blogging process or how everyone who came to our apartments was always disappointed to find that it didn't look like some poverty-stricken hippie den, except for maybe a loaf of stale bread on the counter and some natural products scattered about.

Colin was on his way to a meeting, so we wrapped up our date pretty quickly. He offered to take me through the market and point me in the direction of the Strand, so we paid our bill and started walking. He began asking me if I knew where to find sustainably manufactured running shoes, but what I really wanted to do in my last few minutes with him was shift the conversation back to that e-mail exchange we'd had a while ago about not using toilet paper. He'd written a fairly curt reply about a bowl of water and his hand, but I wanted details—specific, technical details—and now that we'd gotten to know each other better, I thought it might be somewhat more appropriate to press him.

But how could I casually transition from ethical sneakers to bum wiping?

I gave it my best shot, improvising a relaxed demeanor.

"So, Colin . . . are you still . . . uh, you know . . . using your hand? Like, in the bathroom?" I asked as we strolled by an organic fruit stand. Totally smooth.

"Oh, no," he said. "I'm back to toilet paper now."

Aha! So he couldn't have really been that into it. I told him I was currently off TP for number one but couldn't quite convince myself to go all the way. Once again, I tried to bring up my concerns about bacteria, but Colin cut me off with the same succinct response he had before: I just had to get over it—just put my Western squeamishness aside and do it.

"I'm heading this way now," he said, pointing over to his left, "but there's the Strand, down there. See it?"

That was the end of our crap-talk, evidently. I thanked him, hugged him farewell, and said I would drop him an e-mail if I found any eco-friendly running shoes. As I walked toward the bookstore, I wondered if Colin was perhaps being purposely coy. Maybe he planned to go into more depth about his bathroom proceedings when it came to writing his book and didn't want to leak—pardon the pun—any of this information to me beforehand. If this was the case, I was definitely going to be highlighting that page.

DECEMBER 31, DAY 306

Use cold water for washing hands, face, and dishes

Fortunately, I haven't had to wash many dishes recently, thanks to all the Christmas parties and dinner dates. And tonight, I'll be out once again for New Year's Eve: Ian, Dimitris, Jacob, and I—along with a couple of other close friends—are meeting for Indian food at a vegetarian-friendly restaurant within walking distance of my place. Then we're picking up Meghan and heading off to do what we do every New Year's Eve, more or less: go up to Jacob's dad's house in north Toronto (his dad is never really there, so it sits empty most of the time), have a dance party in his living room, eat finger foods, drink fake champagne, and take part in a ritual known as "cake facing." This is something Ian and Jacob do. They go to the local supermarket and purchase a cheap white layer cake with lots of frosting. Then when the moment strikes (i.e., after a few glasses of wine, most often as a song by the Cure starts to play), they plunge their faces into it. They insist it feels liberating; however it usually takes a fair amount of rather nonliberating cleanup afterward, and of course there's the inherent guilt of wasting food. (Although, as Meghan is quick to point out, there's hardly any nutritional value in white cake.) Other than this, we often have a game or activity of some sort—one year, we did an *Iron Chef*–style cook-off where the secret ingredient was a McDonald's hamburger; another year, we played games in the dark, like hide-and-go-seek. A particularly successful occasion was the time we rented fat suits and held a sumo wrestling competition in the basement. This year, we decided to tone it down a little and read short stories to one another, with *Great Gatsby* attire optional. No doubt, this was going to be the geekiest New Year's party in the entire city of Toronto, and none of us would have wanted it any other way.

Despite it being just another night with my old friends, it still counted as a special occasion in my green books, which meant I was officially permitted to blow-dry my hair for a full ten energy-sucking minutes, plug in my hair iron for another fifteen, and apply my toxic but oh-so-volumizing mascara. After a few swoops of blush and dabs of concealer, I put on some of my PeaceKeeper lipstick, kissed a square of recycled toilet paper, turned off all the lights, and skipped out the door.

january

1	Use biodegradable chain lube on my bike
2	No more new plastic
3	Choose subway transfers printed on recycled paper rather than laminated paper
4	Use a soap dish made from recycled chopsticks
5	Collect and return rubber bands from newspaper delivery
6	Choose a real estate agent with green practices
7	Eat only at restaurants that have gone through a green audit and offer sustainable packaging options for take-out
8	Use Coccoina, an all-natural glue
9	Use used boxes for moving day
10	Use unbleached, organic cotton produce bags
11	Frame art with reclaimed barn board and recycled glass
12	Use recycled paper CD sleeves instead of plastic jewel cases
13	Get used parts and tune-ups from a green-minded bike shop
14	Shop at green malls
15	Use only the small stovetop burners
16	Send electronic invitations instead of paper ones
17	Make a bicycle fender from an old water bottle
18	Use a green moving company
19	Eat all the skins on my fruits and vegetables
20	Have my taxes prepared by an eco-friendly accountant
21	Buy organic cotton or bamboo blankets
22	Educate others about green alternatives
23	Learn shorthand to reduce paper use
24	Squeegee shower to prevent mildew
25	Take a butchering class to confront my meat-eating
26	Consume only organic maple syrup
27	Buy a used mattress
28	Choose an eco-friendly tenant
29	Switch to Bullfrog Power, which uses alternative energy sources
30	Use a pumice stone instead of sending pilled sweaters to the dry cleaners
31	Follow maps and get detailed directions for road trips

JANUARY 4, DAY 310

Use a soap dish made from recycled chopsticks

"Close your eyes," said Jacob.

We were sitting in his seafoam-green 1992 Toyota Tercel, the car that most often remains parked in his dad's garage with a flat tire. Now, however, we were outside Ian's house—the three of us were about to go out for dinner to a new restaurant called Citizen. It was the sophomore effort of my favorite Toronto chef whose other restaurant, a cozy bistro called Rosebud (the man has a thing for *Citizen Kane,* it seems), has amazing dairy-free oxtail gnocchi . . . unfortunately this wasn't an option for me now because I had no idea where the meat came from.

"They're closed," I said.

"Okay," said Jacob. "Now, I'm not sure if you're going to like this, but . . . well, here."

I felt him place something flat and square in my hands.

"All right, you can open them now," he said.

I looked down and saw a CD—it looked new, yet somehow dated. With a primary color scheme, the album cover featured a woman in tight jeans and a cropped T-shirt posing as though she were in a weight-loss ad: turned to the side with one knee bent slightly, hands on waist, chin down and looking over the shoulder with a wide grin. Above her was a bunch of writing, in yellow, and I had no idea what it said. This is because it was in Arabic.

"It's Shereen's latest album," said Jacob. "It's Arabo-pop—you know, the stuff we listened to that day when we drove to the Dead Sea and did my recycling."

I did know, and I loved that music. I couldn't understand a word of it, other than *habibi,* which Jacob said meant beloved, but it was so poppy and catchy.

"Oh, yay! Thanks!" I said.

He looked like he may have been blushing, although it was hard to tell in the dark, so I awkwardly leaned over the gearshift, pushed against the strap of my seat belt, and gave him a hug. This was becoming a recurring theme; I'd given the boy at least seven more hugs since first seeing him at Browne's with my parents last week. I could argue that I wanted to make sure he got his fill of them because there isn't much in the way of physical contact in Palestine, but really I just felt like hugging him a lot, so I did, and tried not to overanalyze it.

The next day, I was at Grassroots, my favorite of all the hippie stores where I go to refill my bottles of natural laundry detergent and dish soap. Not only do they have a great bulk section, they also sell some amazing clothes—like dresses made from repurposed thrift store sweaters and bamboo T-shirts—and usually, they have some new product that I can try out for my challenge. It's where I found my biodegradable pen, my recycled photo album, and the plantable greeting card.

As I was looking through the shelves, I saw a soap dish made from reclaimed chopsticks that had been intertwined lengthwise, like the fingers of two people holding hands. It folded up nice and compact, too, which made it good for traveling, and I thought it would be the perfect thank-you-for-the-gift-now-here's-one-back gift for Jacob, especially because he's the type of person who's always careful to wash his hands before he eats and yet not the type of person to put any effort into making sure his soap has somewhere to rest so it doesn't get mucky. It would also make for another green change if I got one for myself because I didn't actually have a dish. So I picked up two.

In all honesty, though, soap—and its accompanying dishes (as well as whether said dishes were eco-friendly in some way)—was probably the least pressing thing on my mind right now. Only two days remained until Jacob went back to the Middle East, and for some reason I was suddenly finding myself aware of every minute leading up to his departure.

This might be why: we'd spent a fair amount of time on New Year's Eve cuddled up on the couch together, listening to everyone read the short stories they'd brought. Then, when Ian and Dimitris and I decided to sleep over, the two of them went off to the guest rooms and I was left to choose between the couch, the basement, and sharing Jacob's bed. Standing in his room with a pair of his sister's old pajamas draped limply over my forearm—there was no way I was sleeping naked anywhere in this house—I bit my lip, looked out the window, and chose the bed.

And so it was that at 3 a.m., on the first day of 2008, my friend of fifteen years and I ended up lying in total darkness, face to face, inhaling each other's exhales. I can't recall much other than these three things: 1) we were talking about happiness; 2) he kept stroking my hair behind my ear; and 3) I was so nervous, I had my eyes shut so hard my sinuses ached.

Nothing happened. Not even a kiss. I had no idea if he even wanted that or if he was just being platonically tender and sweet, but the panic I felt at the mere prospect of doing something I'd regret—having a ridiculous, drunken make-out session screw up a friendship I cherished more than anything—was unbearable. So I kept my eyes closed and my toes clenched, and tried to remind myself that I was just lying in a bed, not dangling over a cliff. Eventually, exhaustion took over and we both fell asleep.

The next morning, Dimitris knocked tentatively at the door before shuffling into the room, sitting down at Jacob's old computer, and immediately launching into a game of Free Rice, waking us up by sporadically yelling out words like "hirsute," "malefic," and "sapience." Our combined vocabulary skills were truly a force to be reckoned with, even at this hour, which meant that we got all the way up to level 49, and, in the process, donated some 3,200 grains of rice to the UN's World Food Program.

Ian stumbled in about half an hour later, confessing to a pounding hangover, and asked if I had any gum to get the taste of stale beer out of his mouth. I said no, but got up to dig around in my purse and finally handed over my mouthwash.

"What's this?" he said, blearily.

"It's Vita-Myr, an all-natural mouthwash," I said. "It's alcohol-free, which is probably a good thing for you right now, and it's got some essential oils in it, I think, like clove and, obviously, myrrh."

"*Myrrh?* You have *myrrh* mouthwash? Like, as in—"

"Frankincense and myrrh, yes, like what the Wise Men brought to Bethlehem," I said. "Just try it. If it's good enough for the baby Jesus, it's good enough for you."

Ian plodded off to the bathroom and returned in less than ten seconds.

"This stuff is horrible!" he said, tossing the bottle back at me, grimacing and shaking his head. "I mean, you *do* realize how ridiculous this is, right?"

I suppose I did, in a way. But on the green continuum of the past year, washing my mouth out with myrrh was as normal as anything, at least compared to sweeping worms off my living room floor and rinsing my hair with vinegar.

"Sastruga!" said Dimitris, suddenly. "Anyone know what *sastruga* means?"

I flopped back into bed next to Jacob.

JANUARY 8, DAY 314

Use Coccoina, an all-natural glue

I'm sniffing glue, quite literally, and none of my brain cells are dying. This Coccoina stuff—another find at Grassroots—is made from pure almond extract and smells amazing. It works, too, which I really don't understand. How could almonds make the back of a glossy photo stick to a page of cardboard so damn well? I'd take the time to research that, of course, but I think I'm actually too high.

JANUARY 10, DAY 316

Use unbleached, organic cotton produce bags

Speaking of almonds, I finally got these organic cotton produce bags I've been wanting for months now. They're perfect for carrying nuts, seeds, raisins, coffee beans, grains, and other small stuff home from the bulk section of the grocery store. I'd tried reusing the flimsy plastic bags that usually come on a roll by all the bins, but they'd get dusty and were annoying to store, and these cotton bags are so much nicer.

I just have to refrain from calling them my nut sacks.

JANUARY 16, DAY 322

Send electronic invitations instead of paper ones

I can't wait for this to end. I really can't. On the one hand, I know I should be reflecting, looking back with Zen-like self-satisfaction at what I've done and acknowledging all that I've accomplished in eleven and a half months' worth of environmental do-gooding, my 322 gifts of green to Mother Nature. But instead, I'm thinking forward. I'm thinking that it's only six more weeks of what is essentially a second full-time job, only a month and a half until I can go for a ride on my sister's motorcycle, order Chinese take-out, pop the cork on a bottle of real champagne, and take a long, hot bath.

So much of this challenge, I know, will stay with me, and many habits are here for good, like using my coffee thermos and avoiding plastic bags. But there are also things I've realized I just don't care about. For instance, as of the first of March, I shall consume avocados recklessly, buy pairs of underwear based not on where they were manufactured but how nice they'll make my butt look, and will both blow-dry and straighten my hair.

But I've made a lot of changes that I'm actually very ambivalent about, too, and I have no idea whether I'll keep up with them. I'm as curious as anyone to find out if I'll still be refilling my shampoo and conditioner bottles next fall, or if I'll continue to avoid toilet paper for number one in another year.

Either way, what I know for sure is that after day 366, there will inevitably be a process of ungreening, of rediscovering the highs and lows of consumption, of balancing my newfound environmental values with my unrelenting, selfish desire for imported cheese, overpriced lingerie, and a warm house in the middle of winter.

The best way to kick off this process, I concluded, would be with a party. A big un-greening party. It could even double as a housewarming because by that time I will have been living in the new house for a month. I'd invite my editor and all my *National Post* colleagues, my old friends from high school and college, my parents and sister, Lloyd from Treehugger—all my social worlds would collide for one night, and one night only.

There'd be bowls of Cheetos in all their MSG glory, gummy bears made from food-grade petroleum, brownies with Reddi-wip, Dutch beer, Australian wine—and it would all be served on plastic plates. I'd ask every guest to drive to my house, preferably in separate cars, circling the block a few times and then leaving them to idle outside. I'd crank up the heat and open the windows, maybe let the shower run and, at midnight, the grand finale: I'd plug in my fridge.

Okay, so maybe I wouldn't go this far, but there'd at least be processed food and a few cases of Heineken. The point would be to let loose for one night, to free myself from the shackles of extreme environmentalism, in as sardonic a way as possible.

Ironically, despite the un-green theme of the party, the invites would have to be sent out while I was still firmly entrenched in my green year, which meant yet another official change: electronic invitations instead of paper ones. Today, then, I logged onto Facebook and began creating an event. It needed a photo, something that conveyed the impression of sticking it to Mother Nature, so I grabbed my camera, went upstairs to my bedroom, unscrewed the shades off one of my bedside lamps, and snapped a picture of me giving the finger to the ultimate symbol of all things modern and green—the compact fluorescent light bulb.

"Love the photo," wrote Matt, from Paris, on the message wall for the event after I'd posted it online. "Can't make it, but I'll bring you a bottle of bubbly a few weeks later." He was coming to visit for Easter.

"March 1st is St. David's Day," wrote my friend Caroline, "but I think I can fit your party in somewhere between my St. David worship sessions."

"Can I bring my infant?" asked Liz. She was involved in a green effort called Project Porchlight, which hands out free CFL bulbs to people in Toronto for their porches. "Infants are VERY unenvironmental, after all."

"Let's have a Styrofoam bonfire!" wrote Josh, a fellow green nerd I'd met through Meghan. "I'll bring my own cup to torch!"

Finally, there was Jacob, to whom I'd sent the e-vite, even though he was all the way back in Ramallah now, hoping maybe he'd see it and say, "What the hell am I doing here anyway? There's an ungreening party I need to be at!" and book the next flight to Toronto (on Estonian Air, via Malta, or whatever).

Instead, he just ticked the "Not Attending" box and sent me a message saying he was sorry he couldn't come, but would throw a recycling bin in the garbage that day as an act of solidarity.

JANUARY 18, DAY 324

Use a green moving company

When I wrote a quick post on my blog this morning about hiring a local, independent moving company for the end of the month, which is when I planned to haul everything from the apartment to the new house—and threw in a pledge to avoid using any tape, packing peanuts, or other environmentally unfriendly material—I thought that would be yet another quick and easy change, and mentally brushed my hands off before heading to work.

By the time I arrived at the office and checked the comments, though, I realized there would be more—so much more—to commit to on moving day.

"Around the corner!!! Truck!!! Yikes," wrote Greenpa.

"Just from the standpoint of least breakage, I'd be tempted to organize a moving party (Free beer! Free pizza! Get 'cher picture on the Internet!) and get about 20 to 30 big strong (male if possible, for your other uses . . .) persons to form a human chain and HAND the stuff around the corner."

My mother was quick to chime in with her trademark sarcasm: "Unfortunately, Vanessa doesn't know 20 strong men (Oh, how I wish she did) and the last move was courtesy of her aged parents! This time they are on strike."

But a bunch of other readers had commented on what a great idea Greenpa had.

"Can you use your bike to move stuff?" wrote blah. "I can just imagine a Radio Flyer being pulled by a bike with a houseplant, a cat and a muffin tin."

I had to admit, that was a pretty cute image.

Then there was a comment from "mtb," whom I figured was Mike the Bike, the guy who recycles bikes and runs a green-minded bike repair shop downtown. He'd fixed my squeaky brake pads a few weeks ago and let me know about his bicycle delivery service, if I ever caved in and ordered Thai food.

"Cargo bike," he said, bluntly. "I have two of them . . . This way you can move your clothes and other small things before the big day. Very, very green."

"Oh Greenpa, you rascal!" wrote just ducky, another faithful Thistle reader. "I think that is a brilliant idea . . . what better way to meet an ecologically minded single young man! Of course, there is the possibility that you might meet a yucko who would then know your address—but where's your sense of adventure?"

Did I really come across as so desperately in need of single, green-minded men?

And more importantly, was I really going to attempt to move everything I owned by foot and cargo bike? This was getting completely ludicrous. Thank god the Centre for Mental Health and Addiction was less than two blocks away.

JANUARY 20, DAY 326

Have my taxes prepared by an eco-friendly accountant

"I need to green my taxes in some way," I said to Dimitris this morning. "Can you do that?"

He'd just started up his own accounting firm, specializing in twenty- and thirty-something artists, the self-employed, and recovering delinquents who tend to keep crumpled receipts in shoeboxes and file their taxes months past the April deadline. He had also secured a much-coveted desk at the Centre for Social Innovation, a supremely sustainable office located in between Toronto's Chinatown and entertainment districts. It had a green roof, was powered by alternative energy sources, designed with eco-friendly materials, had various recycling and no-disposable-cutlery rules in place, and a huge bio-wall in the front foyer. The people who worked here were all involved in nonprofit work or things like eco-school programs and the David Suzuki Foundation, and they often held meetings in a place called the "alterna-boardroom."

"Well, just the fact that I work at CSI has to count for something," said Dimitris.

It was true. I couldn't think of any other accountants working in LEED-certified buildings. Plus, he actually shared his desk with a few people, which deserved another check mark for efficiency.

"There must be tons of ways we can green this," he added. "We can electronically file stuff, to cut back on paper . . ." he paused. "And . . . well, okay, I can't come up with anything right off the bat, but let me think it over for a few minutes and I'll call you later with a whole bunch of ideas."

So I waited.

And he called back.

"Well?" I said.

"Well," he said. "To tell the truth, I can't really come up with anything beyond e-filing. But maybe if you don't staple your receipts together and you ride your bike over to my office, that would count as something extra?"

"Yeah, I'd do that anyway, though," I said. Still, electronic filing and a green office were greener than the paper filing and regular office of the accounting firm I used last year, so I brought down my IKEA storage box—I'd at least upgraded from a shoebox at some point in my mid-twenties—and scratched another day off the green calendar.

JANUARY 23, DAY 329

Learn shorthand to reduce paper use

JANUARY 25, DAY 331

Take a butchering class to confront my meat-eating

Of all the things hippies get worked up about, food is right at the top. No one really gets very uppity about water efficiency, for instance; you don't see intense debates about dual-flush versus composting toilets, and rarely will a friendship end because of a raging fight over who left the tap running. But attempt to extoll the virtues of a cheddar omelette in front of a vegan, and you'd better prepare for a full-scale ethics war.

I've learned this lesson throughout my challenge, which is why my longest posts on the blog are always the ones relating to food. I especially know to walk on eggshells—yes, vegans, *egg*shells—because I've made the decision to continue eating meat, dairy, and eggs. And no matter how hard I try to limit myself to organic, local, grass-fed, hormone-free, non-GMO, free-range animal products, there's just no convincing the noncarnivorous community that this is sustainable.

So when I wrote my post about taking a butchering class . . . well, put it this way: I didn't sleep very well. All I could think about was the inevitable slew of seething comments from readers outraged that I'd even deign to consider this green in any way. I tried my best to explain why I thought it was important to confront my steaks and hamburgers in the flesh and learn how to trim different cuts of meat properly to avoid wasting food, but I knew this logic still wouldn't appease all my veggie detractors. Either way, the class was enlightening and not at all gross—other than perhaps the crackly sound of muscle tearing from muscle—and so I decided that, for better or worse, I'd share with my readers the newfound appreciation I had for the artistry and respect involved in butchering.

After pouring some fair-trade coffee into a mug with leftover tea at the bottom, I sat down and opened my inbox. There were over twenty comments already. Quickly scanning through the first few, I was surprised to see readers leaving words of approval and praise.

"I think this is a great idea," wrote May. "I'm all for us being more deeply connected with our food. Plus, who that grew up reading Laura Ingalls Wilder didn't fantasize about watching animals get slaughtered now and then?"

Wow. A confession of butchering fantasies, even!

"I think it's a great idea you're doing this, too," said blah, one of my regular and almost always supportive readers. "You will really gain an appreciation for what you're eating and thereby be more grateful for the food because you've seen it up close. Good choice."

When I saw a comment from someone with the screen name Raw Vegan Mama, I held my breath, but even she gave her full and unconditional approval: "Everyone just has to do what they can. For me, it is being vegan; for others, it may be eating better meat instead of eliminating it altogether. I applaud you for not ignoring the subject altogether."

But, of course, just as I was thinking I deserved some Nobel Prize in peace and conflict moderation, in came the dissenters.

The first was Derek, who had heard me on CBC radio a little while ago when I was interviewed briefly about my blog for a segment on environmentalism.

"This is total bullshit," he wrote.

Always a good way to start a rational, balanced discussion.

"If you don't go vegan, you're not serious about reducing your footprint [on] the environment, period. Yes, there are more considerations in food choices than just being vegan—obviously eating tofu from China or bananas from Peru while living in Canada is not helping the planet. But avoiding animal products is the most important food choice in this regard! (See UN report.)"

Later in the afternoon, there came another comment critiquing my meat-eating ways, fortunately in a more pertinent and less accusatory fashion.

"I just don't see how you can go for a step-a-day towards 'going green' and still promote eating meat," wrote Dan. "Organically raised, grass-fed—it doesn't change the fact that you're choosing a very green-UNfriendly diet. I don't want to be one of those folks who spreads the guilt, but unless you have a condition that prevents you from giving up meat, it's just as much a wasteful nicety as driving a car that guzzles gas. I just don't see the justification and it really takes away from the poignancy of your efforts."

That one almost hurt more, with the bit about my omnivorous habit taking away from the poignancy of all my other efforts. And he had a point: my justifications for eating meat were definitely more selfish and based on a somewhat falsely bucolic vision of happy animals on happy farms being happily slaughtered.

Nonetheless, I'd devoted hours upon endless hours of thought to this issue, and even if my reasons weren't entirely green or perfectly justified in the end—even if choosing ethically raised meat was really just choosing the lesser evil—it was a decision I was making as conscientiously as possible; and while my value system does, in fact, remain open and malleable, I have yet to be convinced that a single serving of local, organic, free-run meat every week does catastrophic, irreparable harm to the planet.

JANUARY 27, DAY 333

Buy a used mattress

It was the middle of the afternoon and I was lying on an old mattress in the back of a U-Haul office, staring up at the stained, cracked, and blotchy ceiling—or at least what I could see of it in the dark because there were no windows.

A man named Fred was up at reception, dealing with a customer. I'd just met Fred a few minutes ago, and already I was sprawled on his mattress. This was slightly shameful.

"So, do you want it?" came his voice, as he walked into the room and stood against the wall.

I'd initially come to this U-Haul branch with the plan to rent a cargo van so I could drive to the suburbs and pick up a queen-size mattress I'd agreed to buy on Craigslist. The basement of the new house I was moving into had to be rented out to pay the mortgage, and I wanted to rent it furnished, which meant my double mattress would go down there. This, in turn, meant I needed another one, and I figured I'd upgrade to a queen.

But after reading about the dangers associated with off-gas, the airborne chemicals that make up what's generally described as "new car smell" and that can be somewhat toxic if inhaled over a long period of time, I decided not to get a brand-new mattress. Plus, buying something new was never a very green thing to do, so I figured as long as I got a mattress that was clean and firm, whether it was used didn't matter so much.

At first, I'd had a good feeling about the one I'd found at this guy's house up in Richmond Hill—the photos looked fine and when I checked his exact address on Google Maps, I saw that he lived right around the corner from a road called Thistle Drive. That had to be a good sign.

However, while this used mattress in the suburbs technically cost less than a new one—it was $400, but had been slept on only a couple of times in a guest room and had that fancy coiling system that guarantees a sound sleep even if bowling balls are dropping on one side—I hadn't realized just how much the U-Haul rental would cost on top of that. It was only $20 for a cargo van, but there was a separate charge per mile. As Fred pointed out, to drive all the way up to Richmond Hill, back downtown to my apartment, then back up and over to the West End U-Haul office, I'd be traveling enough miles to rack up at least another $100. With tax, I'd probably end up paying a total of $550.

"You know," said Fred, as I began filling out the requisite paperwork, grumbling the entire time about what a rip-off this was, "if you want a really cheap mattress, I've got one in the storage room here that I'm trying to sell. It was at my cottage for years but I'm moving and don't need it anymore. It's a queen size, and I've got a box spring, too. You wanna try it out?"

I did want to try it out. And so there I was. Lying, trying, wondering how much Fred was going to ask for this.

While I pondered, he wandered, pointing to some of other furniture around the room—two seventies-era tables and a love seat that was supremely ugly but in such a cool way—and said he was getting rid of it all, so if I wanted anything else, he'd be happy to give it to me, and would cut me a deal on the van rental, too.

I asked how much he'd charge me altogether—the mattress, box spring, love seat, and cargo van for a couple hours—hoping desperately he'd say a number under $400.

"How about $120?" he said.

$120? Was he serious? Thankfully, before the look of shock could even begin to register on my face, I managed to say the word. Deal.

JANUARY 31, DAY 337

Follow maps and get detailed directions for road trips

Today, I moved.

Although there weren't any more nervous breakdowns, the only reason for this is that I was too stressed out to have one—it would have been yet another thing to deal with, and I didn't have time.

Things began smoothly: I woke up at 9 a.m., got into my moving clothes—loose-waisted pants and a long-sleeved T-shirt—put my hair into a ponytail with an old newspaper elastic, clipped back my bangs, and slid into a grungy pair of sneakers. The whole week leading up to this had been freezing cold with record winds, and there was a blizzard scheduled for tomorrow. Miraculously, however, today was sunny and mild. Well, as mild as it gets in Canada at the end of January.

Everything was packed up, so I went downstairs to make sure Jim, the superintendent, had booked the elevator for me, then returned to my apartment to double-check that everything was arranged and in its proper place.

For the volunteers who had agreed to help, I'd bought some fair-trade hot chocolate and various eco-themed prizes, plus my crank radio was charged and ready to start playing some heavy-lifting music (or at least whatever was on Q107, which at the moment happened to be Pink Floyd's "Dark Side of the Moon"). Only seven people had agreed to volunteer—apparently no one likes to move in the middle of winter, no matter how much hot chocolate is being offered—so instead of handing out maps and directions, I simply told everyone to follow me around the corner, to where my new house was, for the first haul.

The professional guys from Your Friend with a Cube Van were the first to arrive—I still wanted them on hand, just in case, especially for the big stuff like my bed and couch—and after finally agreeing not to use Saran Wrap on the entire couch as a protective mechanism, they immediately got to work bringing stuff down in the elevator. Then came my dad, sister, and Brandon, my sister's friend who lives in my parents' basement. He was studying film and had agreed to record the day's proceedings in order that I might post the video on YouTube for my readers, as proof that I actually went through with this. Soon after, Mason, my colleague at the *Post*, arrived with Lloyd, who planned on writing about this insanity for Treehugger. Finally, there was Mike the Bike and Marianne, the one volunteer I didn't know, who had read my column in the paper the week before and agreed to help for an hour.

By about 4 p.m., we were done. I triple-checked the apartment and slipped the last set of keys underneath the door. There had been only one wipe-out—Mike, on his bike, then off his bike—and only one broken glass. I hadn't eaten anything yet but I could at least relax, so I thanked everyone profusely, awarded Mason the green prize (a tote bag full of recycled scouring pads, a copy of *Plenty* magazine, Burt's Bees products, and a few more eco-trinkets I'd accumulated) and sent them on their way. As my dad, sister, and Brandon began to walk down the street, I went over and stood on my front steps, smiled, and thought about how perfectly my patio furniture would look there instead of crammed on my tiny balcony.

My patio furniture.

Where was my patio furniture?

Oh, yes, I'll tell you where it was. It was with all the other stuff in my two storage lockers, which were in the bowels of the parking garage in my old building. I'd somehow forgotten all of that—not just the wicker chairs and table but my old bike, the extra shelves for my bookcase, a lamp, boxes of photo albums, sporting gear, an indoor grill, and a record player.

I wanted to cry, but again, just didn't have time.

Instead, I ran back, called my superintendent, and begged him to let me in so I could get to the storage lockers, then phoned my dad and put on my best take-pity-on-your-daughter voice.

"You're going to hate me," I said.

"Uh-oh, what is it?"

"I'm not done moving. There's stuff in the storage lockers I forgot and I really, really, *really* need you guys to help me with one—or, well, maybe two last loads."

He said he'd return with Emma and Brandon but they were just grabbing a bite to eat so it would be another half-hour.

And so I waited. But I was forced to stay in the parking garage because if I left the building, I'd be locked out again, and I couldn't even get up to the lobby without an entry fob for the stairs and elevator. On top of this, my cell phone didn't get any reception unless I stood up by the automatic doors, near ground level. Twenty minutes later, then, and the lovely Miss Green as a Thistle could be found crouching on the dirty cement floor of a parking garage, one hand propping up her dejected head and the other holding a pink cell phone aloft in the air, hoping it wouldn't lose the last of its signal strength and also hoping no one was trying to steal her bike and patio furniture, which had been left sitting in one of the empty parking spaces two levels down.

As it turned out, Jacob had tried to call during this time but wasn't able to get through. He'd left a sweetly long-winded message, though, saying he hoped my green move went well and that he was thinking nonstressful thoughts for me, halfway across the globe. I listened to his message three times.

By midnight, I finally had everything moved into the new house, for real this time. I opened my laptop and brought up my music folder, clicking on a Bob Dylan album, then decided to break my green rules about delivered food, organic dairy, and ethically raised meat by ordering a pepperoni pizza because the effort it would take to dial ten digits was truly all I could muster at this point.

After hanging up, I collapsed on Frank's mattress and closed my eyes. Some minutes later, Bob started singing "Tangled Up in Blue." I let out a deep sigh and decided to play this game called Three Wishes with myself—if I could have any three wishes granted right this second, what would they be?

The first was easy: to be finished with this green challenge. I felt, at this point in the game, that I'd learned what it means to be an environmentalist, to live purposefully and responsibly, respecting the earth with a heightened awareness of how its resources are drained every single time a toilet gets flushed, a light turns on, or a newspaper arrives at the doorstep. I still loved making an effort to be green, I still loved being in constant communication with all my readers—both the cheerleaders and the critics—but the effort it was now taking me to juggle 337 green balls in front of an ever-growing audience was wearing me down, and it didn't help that the balls kept getting heavier. The action-based changes—using vinegar to clean my kettle or making organic peach jam—are ones I could easily keep up forever. But the restrictions—limiting my use of water, cutting back on heat, confining my diet to food grown within a certain diameter—are starting to have more of a straitjacket effect. Right now, I honestly can't fathom making one more change to my life in the name of environmental stewardship; conversely, I feel as though the environment is walking all over me.

The second was also easy: to have Jacob here. We'd been writing to each other every day since he'd returned to the West Bank, and in an especially head-spinning e-mail a little while ago, he'd opened up about his feelings for me. I spent an entire day at work in full panic mode, not knowing what to do or how I felt, but the more we wrote and the more we talked, the more I realized how strongly I felt for him, too. I'd been searching for my soulmate for over a year now, worrying that I'd never find anyone who was into both environmental do-gooding and acerbic banter, who'd get along with my parents but ultimately be on my side, and who worked with—not against—my innumerable neuroses. All this time, he might have been standing right under my nose (or, you know, the nose I keep in Palestine), and I let him go with a dozen hugs, yet not a single kiss.

The third . . . what would be my third wish?

I paused and listened to the lyrics.

> The only thing I knew how to do
> Was to keep on keepin' on like a bird that flew,
> Tangled up in blue.

The man was right. This may be an overwhelmingly stressful day, and it's only a small part of an anxiety-plagued year, but all I really needed to do, as he said, was keep on keepin' on. I wasn't so much a bird tangled up in blue—more like a girl tangled up in green—but whatever the metaphor, a sense of peacefulness began to flow through me knowing that at the end of this day, and every day, what I was doing was good, what lay ahead of me could only get better, and if I just kept flying forward with all my strength, everything would be okay.

Just then, the door rang. Pizza.

I tried to think of a third wish, but my brain was focused on my stomach and my stomach was whining in protest, so I went with the old standby—world peace—and bolted down the stairs.

february

1	Shovel snow by hand and use sand instead of salt
2	Install a dual-flush toilet
3	Use low VOC or water-based paints
4	Use a rain barrel to collect water for the garden
5	No more makeup
6	Use leftover kindling and eco-logs in the fireplace
7	Run only one application at a time on my computer
8	Restrict diet to food grown within Ontario
9	Use rechargeable batteries
10	Close curtains at night to insulate house
11	Minimize power usage during peak hours and sign up with Peaksaver program
12	Purchase only recycled glass
13	Fly direct
14	Use lemon, vinegar, and olive oil to polish wood
15	Keep air clean without a plug-in air purifier
16	Tuck my pant legs into my socks to keep pants clean
17	Buy recycled wallpaper
18	Use Green Maps; go on eco-tours
19	Skinny dip
20	Help push stuck cars out of the snow
21	Use toothpaste rather than plaster to fill holes in the wall
22	Provide for an eco-friendly funeral in my will
23	Use lickable stamps instead of sticker-based ones
24	Write poetry in haiku form only
25	Recycle my old running shoes
26	Delete all spam and old e-mails from inbox
27	Use only fair-trade vanilla
28	Fix other people's green mistakes
29	Go to sleep

FEBRUARY 6, DAY 343

Use leftover kindling and eco-logs in the fireplace

After not having cable for months, I've actually forgotten what it was I loved so much about television in the first place. I have vague memories of watching tastefully dressed women flit about their studio kitchens and gesticulating with garlic pressers, of Tyra Banks holding seven headshots as she says solemnly, "Eight beautiful women stand before me," and of obese people in red and blue T-shirts voting someone off their team for not losing enough weight that episode. None of this strikes me as entertaining now.

While at a friend's cottage this weekend—this is a friend, by the way, who happens to have satellite TV with over nine hundred channels—I tried with concerted effort to rediscover my addiction. But as I clicked and clicked and clicked my way through all the shows, nothing other than the news really appealed to me, and I get that from the paper each morning anyway. But just as I was about to give up hope, I saw it. My saving grace. The light at the end of the prime-time tunnel. It wasn't *Entertainment Tonight*, nor was it a rerun of *The Office*. No, it was something much greater—it was *The Fire Log Channel.*

Now this, I thought—this was something.

Essentially just a single continuous take of a log burning in a fireplace, the concept was so brilliantly ludicrous, so high-tech-Luddite, and oh, so ironic, I was immediately transfixed and sat there watching it for a full twenty minutes.

In this time, I wondered:

Who's responsible for this hearth?

Is anyone stoking this fire?

What are the production values for *The Fire Log Channel*?

What kind of wood is being used?

Did they have to audition logs?

Are there subliminal messages being conveyed through the flame patterns?

Who's the target demographic here? Lumberjacks? Pyromaniacs? Lumbermaniacs?

Will there be another season? A blow-out *Fire Log* series finale?

In fact, I wasn't too far off in wondering if there would be further episodes in the show—according to the network, audience demand for the *Fire Log* was so great, producers had no choice but to bring it back, extending its season to the end of February and adding a second channel, so viewers could catch the action on either 238 or 299.

As the on-screen TV guide so seductively argues, "it only takes a second to ignite your glowing fire."

Indeed, I thought—consider me lit.

Eventually, however, I realized that while the televised fire did give off trace amounts of heat from the screen, it wasn't enough to actually keep me warm and so I decided to build a real fire. Although there are "green" Duraflame logs with natural wax instead of petroleum, which the company insists burn cleaner than wood, as well as eco-friendly logs made from old coffee grounds, the old-fashioned Canadian in me was stubbornly convinced that scavenging for whatever dry branches were lying on the ground outside still remained the most environmentally sound option.

Of course, trying to actually build a fire and sustain it is a whole other challenge. Again, along with "How Your House Works," "How to Make a Fire" should really be a mandatory component of the modern-day education system.

FEBRUARY 8, DAY 345

Restrict diet to food grown within Ontario

If I have to eat one more beet, I will kill somebody.

FEBRUARY 13, DAY 350

Fly direct

Jacob and I have decided to meet up in Spain for our first date. Considering we're already talking every day, exchanging Keats poems, and saying "I love you," it only makes sense that we at least try to kiss each other—after all, there's still a chance that we've drastically overestimated where we belong on the romantic-platonic continuum. Better to sort this out sooner rather than later.

We were originally going to choose somewhere more obscure to meet up—I was voting for Montenegro, he made a case for either Estonia or Lebanon—but despite my being unencumbered by green restrictions in April, it nonetheless seemed a bit excessive to travel to a destination requiring multiple flights and car rentals, not to mention border crossing and political tension; nothing kills the mood, after all, like a military coup. So in the end, we settled on somewhere halfway between Toronto and the Middle East and booked two direct-as-possible flights to Malaga, in the south of Spain. I'm hopeful it will also be a direct flight to a seamless and most unawkward weeklong date with one of my best friends.

FEBRUARY 14, DAY 351

Use lemon, vinegar, and olive oil to polish wood

In the way that people don't realize how pale they look until they put on a swimsuit after a long, dark winter and hit the beach in the middle of the day, I've noticed that I actually don't recognize how green I've become until I put myself in a normal situation. Despite all my hand-wringing over the occasional bagel with nonorganic cream cheese or extra thirty seconds with a blow dryer, my greenness stands out very starkly as soon as I leave the house.

This past month, for example, I'd been getting what I thought were hot flashes and was starting to panic that I was hitting an early menopause. But then it occurred to me that I felt these waves of heat only in other people's houses, especially my parents' house, where the thermostat is cranked up to almost 82 degrees Fahrenheit in order to accommodate my mother's poor circulation as well as compensate for her having to grow up in a freezing cold house in the perpetually damp and overcast British Midlands.

I had been taking part in Crunchy Chicken's latest challenge, Freeze Yer Buns, in which all her readers were turning down our thermostats as far as we could handle. Mine was set at 64 degrees, and while my body had gradually adjusted to the cooler temperature, I now felt uncomfortably hot in most other indoor environments.

Alongside my initial concerns about the hot flashes was also a worry that my allergies were getting worse despite it being winter, when there shouldn't be much more than dust in the air—certainly not ragweed, my official arch nemesis in all things sinus-related, as that wouldn't crop up until August.

Again, though, I realized that my nose itched and my eyes watered only when I was at other people's houses. Well, not every house—Meghan's apartment, for instance, was always fine, probably because she uses the same products I do and is careful about eliminating any and all forms of toxins from her living space. But when I was staying over at another friend's place and had to use her concentrated, Clean Breeze–scented, neon green laundry detergent as well as the purple lavender dish soap, both of which were crammed full of artificial fragrances, my eyes kept bursting into tears and my nose suffered perpetual seizures.

I've always prided myself on not being one of those flaky, ultrasensitive types with weak immune systems. But after making my body adapt to a more natural lifestyle, it's apparently decided that, from now on, it will accept nothing less—at least not without getting all hot and bothered and maybe crying—which is why I've now been marinating the armrests of Fred's ugly-awesome loveseat in a homemade blend of lemon, olive oil, and vinegar, instead of buying an aerosol can of Pledge.

FEBRUARY 17, DAY 354

Buy recycled wallpaper

Supposedly, Oscar Wilde's dying words as he lay in his hospital bed were "Either the wallpaper goes or I do."

Now I'm sitting in the hospital, visiting my friend Kieran and telling him all about the recycled wallpaper I just bought for my home office, and how imperative it is that he survives so he can see it. At the very least, it would be nicer to look at than jaundiced beige paint, a faded curtain, and another faded curtain, which looked as though they'd been purchased for $12 each at that warehouse where I got my old hotel chairs.

In fact, from both an aesthetic and an environmental perspective, this hospital was all wrong. The color scheme was muted neutrals and lifeless pastels, there didn't seem to be recycling bins anywhere, and the food—well, everyone knows the only stuff worse than airplane food is hospital food, but considering this was the ward for patients suffering from Crohn's, colitis, and other bowel conditions, one might think there'd be something on the menu slightly more nutritious than green Jell-O and yellow Gatorade.

"Is that a bagel wrapper?" I asked him, with an audible *tsk*. "And a can of pop?"

If Meghan were here, she'd be in spasms of disapproval.

"Yeah, well, I needed carbs and it was better than what the cafeteria was serving," he said. "And that's flat ginger ale, by the way—supposed to be good for the stomach."

He said he couldn't eat many vegetables right now anyway and figured he should listen to his cravings in order to put some weight back on. As long as it stayed in his system, he was happy. Still, I just couldn't believe food coloring and refined sugar could possibly be good for anyone, let alone someone with severe digestive problems.

"Yellow Gatorade is my favorite, too," he said, smiling. After swallowing his last sip, he let it drop into the garbage bin by his bed.

I remembered the conversation with my mother about all the waste created by the medical industry and hospitals—the disposable syringes and paper gowns were one thing, but it just seemed ridiculous to not even recycle cans and bottles. If hospitals made the effort to do this, and maybe fed the patients some decent food, this could translate into cleaner air, water, and land, not to mention healthier people, which in turn might put less strain on the health care system in the first place.

The nurse came in to change Kieran's saline drip and take his blood pressure. Shortly after, a voice came over the speaker system announcing visiting hours were over.

I reached into the garbage and plucked out his empty bottle.

"Mind if I recycle this for you?" I asked.

Kieran gave me a look that suggested he was either impressed or grossed out, or possibly both. "It's all yours," he said. "Shortbread cookie for the road?"

FEBRUARY 18, DAY 355

Use Green Maps; go on eco-tours

I would never have thought to green my cartography, but as I'm often reminded, despite all evidence to the contrary, there are even nerdier environmentalists out there than me.

Walking out of a garden store this afternoon, a map caught my eye—it looked to be green in some way (I now have razor-sharp peripheral vision when it comes to spotting anything with enviro-themed design), and sure enough, it turned out to be a map of Toronto that highlighted all parks, ravines, and bike paths in the city, as well as every sustainable hotel, eco-friendly dry cleaner, vegetarian restaurant, community garden, farmers' market, and even all the different recycling organizations.

The company behind them is called Green Map, I discovered, and they've done similar work for cities such as Stockholm, Chicago, and Beijing (this one confused me a little—were there really green parts of Beijing?). The funniest, however, had to be the compost map of Manhattan—I mean, how many banana peel drop-off locations do New Yorkers really need?

FEBRUARY 22, DAY 359

Provide for an eco-friendly funeral in my will

As my challenge nears its final resting place—and, more matter-of-factly, as my lawyer keeps reminding me to write a will, what with the house ownership and all—the subject of death seems only appropriate. Although I don't plan on decreasing the population by one anytime soon, I am prepared to green my mortality.

I remembered Crunchy Chicken addressing this topic before on her blog, so I dug up the old post on it and read through the comments. After some further, albeit rather inconclusive research, it seemed that, ultimately, the most environmentally sound route into the afterlife was either burial in an eco-cemetery with no accoutrements or cremation in a pine, cardboard, or biodegradable box, providing my teeth didn't have fillings, which they didn't (the mercury gets released into the air).

But there were even more options to choose from if I wished, such as a sky burial, which would mean I'd be dragged up on a mountain for vultures to devour. Or there were these things called Ecopods, coffins made from naturally hardened recycled paper. There was also the "memorial reef ball" idea, in which my ashes could be mixed into a concretelike substance, molded into a dome structure, and dropped to the bottom of the ocean where it would cultivate coral and help develop aquatic life. And, of course, I could always donate my corpse to science for medical research—or even creepier, give it to the team behind the touring Body Worlds exhibit, who'd most likely flay me and pose my remains in the form of a tennis player, midbackhand (or more accurately, an environmentalist mid–compost churn).

In the end, though, the idea I found to be most sustainable and moving was that of getting cremated and having the ashes poured in a compostable urn with some seeds, so that it could be planted; then, in the future, a tree might grow in its place. There's even a company that manufactures something called the Bios Urn, which breaks down in the earth gradually. Its logo, rather disturbingly, looks like a person's head locked within a recycling symbol, but otherwise the product appears trustworthy.

This means no formaldehyde or embalming fluid, no need to

manufacture or transport a coffin, and there aren't any cemetery fees or maintenance issues such as pesticide and motorized lawn mowers. On top of this, I think I may also write that I'd prefer donations to an environmental cause in lieu of flowers and, if possible, no hearse—unless it's a hybrid hearse. Or maybe it could be a bicycle funeral! Ooh, and it would need to be carbon-offset, so it could be a no-impact ceremony!

Who knew planning a funeral could be so much fun? I was definitely going to rest in green peace now.

FEBRUARY 24, DAY 361

Write poetry in haiku form only

Although this is probably the most questionably eco-friendly change I've made so far—writing poetry in haiku form only—it at least has to count as the most creative and romantic change, too.

I've never written much poetry, but what I appreciate about the haiku is that it forces you to speak your mind in seventeen syllables—no more, no less. This means there's enough space to say what's important without droning on effusively about it. And it's the most environmentally friendly form of poetry because it's so short and therefore requires less paper and ink.

For my blog post today, I decided to write a haiku for my readers, one that hinted at the possibility that I'd be keeping up my website after this challenge was over. I didn't know how I was going to do this or what I'd write about, but it didn't really matter because the restrictions of haiku meant I pretty much had to keep things vague.

Ahem:

Being Green
by Vanessa Farquharson

My year is ending
but there's so much more to come;
stay tuned, dear readers.

That was it. Not particularly moving or thought-provoking, but it would do. A few hours later, I checked my blog again and found that a bunch of readers had left comments in haiku form.

Hellcat13 wrote:

Me and my friend Er
Write haikus to each other
When we're bored at work

Then Esme threw another out:

Find 'green' sunglasses
Donate to Heifer.org
Two green haiku thoughts!

I was inspired to keep up the verse, so after a few minutes' thought and counting with my fingers to double-check the syllables, I threw in another:

Haha, I love it!
Can we get all comments in
haiku form only?

Sure enough, more came in. Tuuli wrote:

I love this new post
I featured you on my blog
Because you inspire

Then David ben-Avram wrote:

Thank you for the blog
I love poetry and you
So hot and so green!

Er, all right, that last one was a little too GreenSingles.com for me. Suddenly, my interest in seventeen-syllable blog comments was waning.

After I closed my browser, I leaned back in my chair and stared up at the ceiling, wondering if I should bother changing the halogen light fixtures that came with the house to accommodate compact fluorescents or LEDs. Just then, my friends Joel and Amy wandered upstairs to say they were leaving.

For the past couple of weeks, the two of them had been staying in my basement, helping me fulfill my rule about sharing living space. They both teach English at a primary school in Jeju, in Korea, but were home in Toronto for a visit and didn't have anywhere downtown to stay. I'd offered my house, seeing as I was currently the only one there. Realizing this was the climax of my green challenge, they were more than happy to give me some space and abide by most of my eco-restrictions. However, while Amy used to be a pretty hardcore hippie in her early twenties and is thus familiar with things like Dr. Bronner's organic soap and the bulk section at Grassroots, Joel is far less concerned with the environment—in fact, he is less concerned with most things that don't involve eating, drinking, movies, music, and books, which is why I love him.

After a solid fortnight of running around socializing, shopping, and brunching, Joel and Amy were finally ready to fly back to Korea. I went downstairs to say goodbye, then spent the next little while absorbing the stillness.

Around noon, the doorbell rang.

It was a deliveryman with a bouquet of yellow orchids. The card read, "Thank you so much for letting us stay in your home. These flowers are fair-trade and the packaging is all biodegradable. Hope you like them, and best of luck finishing your green challenge—let us know how the party goes!"

That was in Amy's handwriting. Underneath, Joel had scrawled something else.

"P.S. Fuck the environment."

Green, but cynical. They knew me well.

FEBRUARY 28, DAY 365

Fix other people's green mistakes

For so many months, I'd been patiently waiting for the moment when I could type "Day 365" into the subject heading on my blog. But thanks to the evils of the vernal equinox and all of this leap year business, I've been forced to come up with one extra change. So perhaps it was due to some lingering resentment that I ended up with today's post about fixing other people's green mistakes—moving plastic bottles from the garbage into the recycling bins and moving polystyrene containers in the other direction, turning off a tap when the person next to me is letting it run for no reason, asking people to stop idling their cars, and so forth.

Except the frustrating thing is, even with 365 green changes under my belt, I'm still far from being an expert on anything environmental. I've been asked to speak at a few upcoming Earth Day events and had to warn the organizers that I don't actually know what I'm talking about beyond how disgusting used handkerchiefs are or what grass-fed burgers taste like. As well, more and more friends have started asking my opinion on things like cloth diapers versus Seventh Generation disposable diapers, or have had questions about whether cardboard boxes need to be collapsed before going to the curb for recycling and if greasy pizza boxes can be recycled at all. I still don't know the answers to these things.

Clearly, the infrastructure of this planet is far more complex and nuanced than any individual can grasp (except maybe Stephen Hawking). This is probably why we're still debating whether ethanol gas or electric cars are the way of the future, arguing about the merits of solar versus wind power, and even commissioning studies to figure out precisely to what extent global warming is man-made. The difficult thing with the green movement is that most average citizens living in first-world countries, like myself, are going to have to make a lot of changes to their way of living without fully comprehending why, beyond the catch-all phrase "It's good for the environment." Sometimes, we'll even have to make a few wrong changes first—one day, for instance, PET plastic wine bottles will be declared the greenest on account of their lightweight recyclable packing; the next day, it'll be Tetra-Paks because of the hormones leached in PET; and eventually it'll come back to glass because there's no oil left to make plastic. Does this mean we shouldn't bother making any changes because critics and scientists keep changing their minds? Of course not. But I do wish it were possible to get a better grasp on which ones have the biggest impact, rather than taking 366 stabs in the dark and hoping for the best.

FEBRUARY 29, DAY 366

Go to sleep

At 10:30 a.m. on the last day of my challenge, I could be found leisurely sipping some fair-trade coffee from my thermos, watching Tim Roth get bludgeoned in the knees with a golf club by Michael Pitt while Naomi Watts, bound and gagged in nothing but her underwear, sat crying hysterically behind the couch. It was a press screening of the film *Funny Games*, an intense and controversial thriller by Austrian director Michael Haneke, and I was at the same theater I was at most Friday mornings, where all the upcoming films are screened for the press so we can write our reviews and choose our star ratings ahead of time.

It was as average a Friday morning as any, yet I felt especially thrilled—not from anything that was transpiring on the screen, but rather from the sense of accomplishment swirling inside my chest. In just over twelve hours, I'd have done something no one else had done. I'd be able to look back at 366 changes I'd made to my life for the sake of this earth, let out a sigh of exhausted pride, and eat, eat, eat—I could eat anything at all, like bananas and Cheetos and sushi and guacamole, and then I could run a hot bath and stay in it until I was too drunk on warmth to climb out.

I had been wondering for a while now what I should do for

my final change—whatever it was, I'd have to do it for no more than twenty-four hours. This led to thoughts of going all out: not showering, not eating, not drinking, not using anything, not doing anything, not going anywhere or buying anything—maybe even slowing my breath and carbon-offsetting whatever CO_2 I did emit in a day's worth of exhaling. But something about this felt too gimmicky, like I was just skewering the whole concept of extreme green living when, in reality, I do take it seriously.

So when it came time to make my last pledge and I still wasn't sure what to do, I did what I learned at the Banff Centre, a classic writer's block technique: I went back to the beginning.

Recycled paper towels.

I began this whole thing with recycled paper towels.

I'd come a long way since then—giving up paper towels altogether, in fact—and my initial goal of making small steps toward environmentalism had completely disintegrated upon realizing there were only so many steps I could take before leaps and bounds became the only option.

But while I'd changed my ultimate philosophy from believing the path to a better planet lay in lots of minor, not major, lifestyle changes to the opinion that both big and small changes were necessary, I nonetheless felt it was important to conclude with something simple.

That something, in the end, was sleep.

The idea came from one of my readers, Molly, who in the comments section of my post on haikus had copied a link to a story in the *Guardian* about how sleeping is good for the environment. It was overly conceptual, probably littered with speculation and more than a bit tongue-in-cheek—right up my alley. Despite the initial silliness of its claims, the article made some valid points about how an earlier bedtime meant that lights got turned off sooner and appliances got used less, as well as the tendency for well-rested individuals to use fewer resources than those who are stressed.

Green as a Thistle began small and it would end small, too—it would end with me, and the blog, taking a long nap.

When I got home from the movie, I started cleaning the house and getting everything together for the big un-greening party that night. Ultimately, I just couldn't bring myself to buy Styrofoam plates or plastic cutlery, so a bag of Cheetos and some Australian shiraz was as far as I'd plunge the dagger into Gaia's chest. After setting these out on the living room table, I made my way upstairs, sat down at the computer, and checked to see how many comments people had left on my final post. There were eighty-nine.

It was almost hard to read all of them—I've never been good at accepting feedback, whether criticism or praise, so eighty-nine offers of congrats and kudos, many of which detailed how I'd inspired someone to make a difference in his or her own life, really shocked me into a state of humble disbelief. Clearly that cliché of the ripple effect actually had a lot of truth to it.

But it was the ripples closest to me that meant the most—over the course of a year, I watched my friends and family make changes I never thought they would; at first, it would often be for my sake, just to accommodate my green restrictions, but now I truly believe they're doing it for themselves and for the earth. My mother's fridge is now only ever stocked with organic dairy and free-range meat, and she's sworn that every Christmas she's giving donations to her friends instead of presents. My father is renting only subcompact hybrid cars when he goes on vacation, and he always drinks organic beer. Even Emma is . . . well, Emma hasn't really changed, actually.

But Ian! Ian is bringing a coffee thermos to work every day and is even planning on voting green in the next election, and Dimitris has bought a used bike to cut back on his driving. Even my colleagues at the *Post* have started drinking water from reusable stainless steel bottles and packing homemade lunches.

My friends, themselves, have also changed—a couple of years ago, I never would have predicted I'd be flirting with Toronto's solid waste crew, having coffee with a guy called No Impact Man, or dating a vegan massage therapist from Oregon. Nor, for that matter, would I ever have found myself tracking down recycling cages in the West Bank, pre-ordering worms for a compost bin on my balcony, switching tampons for a bendable cup every month, or trawling the aisles of a warehouse full of discarded hotel furniture before rushing home to sterilize my Mason jars and make jam.

I don't know whether these experiences have made me a better person—if anything, my ongoing spiels about the miracles of baking soda and vinegar have probably rendered me annoying to some—but they have made me a more responsible, less insecure person. There's this belief that having too much knowledge can be a burden, that it can lead to overwhelming frustration at the ignorance of mankind, or to depression over the inability to change things. To a certain extent, I can understand this—the facts about global warming aren't going to get any easier to digest, and the endless jargon, conflicting studies, and exceptions to the rule just make it worse. But learning that shoes can be polished with coconut oil, showers can be taken in the dark, vinegar can clean the calcium buildup at the bottom of a kettle—all this gives me immense pleasure. Much like when Dorothy in the *Wizard of Oz* is told that she actually had the power to go home the whole time, my challenge has taught me that everything I truly need in life is right here already: cleaning products, cough remedies, and all the ingredients for toothpaste are sitting downstairs in my pantry; exercise equipment is outside in the form of hills and wind; my mode of transportation is my own two legs; and, as much as I love a good mascara, I can also make myself look pretty by smiling a little more.

All this is courtesy of Mother Nature, and it will continue to be here for years to come, providing I show her some love in return. This reassures me. Before, I'd always get worked up about things that needed doing, stores that needed going to, people who needed to be seen, products that needed purchasing, menus that needed planning, and so on. Now, however, I know that an empty fridge doesn't translate into an empty stomach, nor does an empty wallet mean an empty closet—in fact, the less stuff I have in my life, the more fulfilled I feel, which sounds Buddhist and hokey, but it's true. Even when it comes to my love life, it seems the answer to that might have been here all along, too.

With all these thoughts gathered in my head, I walked into my bedroom and lay down. It really had come full circle now; after all, I'd had the idea for this challenge while lying in bed one night, just over a year ago, trying without success to fall asleep. I couldn't fall asleep now, either, but it wasn't because of any environmental guilt or restlessness. No, this time it was pure giddiness that came over me. I listened to the bare tree branches outside sweeping at the roof, and the muffled, rhythmic scraping of a neighbor shoveling snow. Rolling over toward the window, I watched the cold winter light valiantly punch its way through a cloud. I felt aware of the world in a beautiful way, not an anxious one, and I smiled.

epilogue

To my utter shock and amazement, the planet didn't implode, melt, or stop spinning as soon as I finished my green challenge. It was almost disappointing. For those first few weeks of freedom in March, I couldn't help but feel uneasy every time I took a hot shower, poured a messy stir-fry onto a clean plate, or bought blackberries from Mexico, despite the fact that there were never any repercussions, at least not that I could see.

The party was a big success, although I immediately unplugged my fridge once again the next morning, and while taking out an entire garbage bag full of disposable plastic cups, I felt absolutely sick with guilt (then again, that could've been the Cheetos lingering). In short, while certain sins were easy to recommit—using hair dryers, drinking Beaujolais, shopping at H&M—a lot of my changes were proving difficult to unchange.

Fast-forward another eight months: at this point, a lot of people had been asking how much of the green stuff I was keeping up, so I decided to return to my blog and the master list to scroll through everything and get an official number. Some of the things weren't applicable now that I was in a house instead of an apartment, and some later items on the list canceled out earlier ones, but the figure I eventually came up with is 271 (about 74 percent). That's a lot. Certainly, it doesn't feel like I'm keeping up 271 changes, but then again, to paraphrase a common saying, once you go green, you never go back. Now that I'm buying natural products that are just as effective as any conventional brand, there's no reason to revert to what I used before; now that I've made the effort to get a rain barrel, I'm not going to get rid of it; and now that I've started using glass jars instead of plastic to hold my food, I don't see the point in tossing those and buying regular Tupperware all over again.

Oh, and I'm definitely not buying a car.

Even on the more behavioral side of things, I continue to use my water-bottle-and-washcloth system instead of toilet paper for number one; I try to bring my own containers when getting takeout at a restaurant and almost always enforce the rules about ethically farmed meat; I refill all my bottles, still don't have cable, order ice cream in a cone, and fill my kettle every morning only with the exact amount of water needed.

What do I not do? Well, I don't pick up other people's litter or let it mellow; I use my dishwasher, my oven, and my blessed vacuum cleaner regularly; I shave my legs, drink wine from a glass, and, unfortunately, I still haven't perfected my shorthand. I've also toned down my THESE COME FROM TREES sticker campaign, but this is mostly because all the cinemas in Toronto have switched over to unbleached recycled paper towels in their restrooms.

But going back to the numbers for a moment—what constantly irked me during the challenge was that I didn't know *exactly* how many trees I was saving, or *precisely* how much carbon I'd prevented from going into the air. So I decided to ask the folks at Zerofootprint, a local offsetting and environmental auditing company, to look over my changes and see if any hard numbers could be deduced. Recently, they got back to me, and while it was impossible to tabulate things such as signing up with Freecycle or using organic conditioner, they were able to calculate an official number for ninety-five of my actions (that's about 26 percent of the total), and it came to 11.02 tons of CO_2 saved. Not bad.

In terms of keeping green with regards to the blog and my column at the *Post*, I have to admit the entries at Green as a Thistle have become somewhat infrequent; however, I'm writing a feature column called "Sense & Sustainability" every week for the paper, which means I get to spend my work hours learning about all the amazing green changes happening around the world, such as the young British guys setting up a permaculture farm in Palestine, the executive VP of a Toronto hospital installing solar panels and trying to minimize waste in our medical system, and a group of artists in California assembling a global resource of "feral fruit maps," promoting free access to berry bushes and fruit trees growing in public domain. This stuff makes me happy.

Oh, and speaking of happiness, there's Jacob.

In early April, we made it to Spain. Our first kiss was in the dingy, crowded arrivals hall at the Malaga airport and it was perfect — not a trace of awkwardness. We drove out in a fuel-efficient, subcompact car to an adorable villa in the Andalusian countryside and spent the week talking, swooning, eating, laughing, and cuddling. I drew him a picture of the lavender bushes by the pool while he wrote my name in Arabic; later I stared into his ear while he sang me a lullaby. Every night, we sat by the fire, listening to the bells tinkle on a herd of goats in the distance.

Let me just say: there is nothing quite like falling in love with one of your best friends.

Jacob is still working in the West Bank but will be coming home to Toronto in a couple of months and moving in with me. This means at least two of my green changes will continue to be in effect: sharing my living space, and sleeping naked.

acknowledgments

There are too many people to thank for this, but let me try to be quick and keep it at ten: 1) my family; 2) my pseudo-anonymous readers, who gave me so much sage advice; 3) Meghan, for supporting this venture from the get-go and feeding me organic lentils when I was down; 4) all my friends who put up with constant demands for vegetarian restaurants and Ontario wine; 5) my editors at the *Post*, Ben and Maryam, who for some reason kept me on the payroll despite numerous days of absence in the name of book writing and environmental do-gooding; 6) my agent, Sam Hiyate, for the martinis and ego-boosting; 7) my editors, Robert and Lisa, at Wiley and Houghton Mifflin Harcourt, respectively, for believing me when I said this could maybe be a green book that people would actually want to read; 8) Ms. Carrier, for giving me the Grade 11 English award, not to mention some semblance of writerly confidence; 9) Sarah K., for name-dropping me to Margaret Atwood; and 10) Jacob, for knowing they were hazel.

acknowledgments

[illegible]